GCSE OCR 21st Century
Chemistry
The Revision Guide

This book is for anyone doing **GCSE OCR 21st Century Chemistry** at higher level.

GCSE Science is all about **understanding how science works**.
And not only that — understanding it well enough to be able to **question** what you hear on TV and read in the papers.

But you can't do that without a fair chunk of **background knowledge**. Hmm, tricky.

Happily this CGP book includes all the **science facts** you need to learn, and shows you how they work in the **real world**. And in true CGP style, we've explained it all as **clearly and concisely** as possible.

It's also got some daft bits in to try and make the whole experience at least vaguely entertaining for you.

What CGP is all about

Our sole aim here at CGP is to produce the highest quality books — carefully written, immaculately presented and dangerously close to being funny.

Then we work our socks off to get them out to you — at the cheapest possible prices.

Contents

How Science Works

Module C1 — Air Quality

Module C2 — Material Choices

Module C3 — Food Matters

Module C4 — Chemical Patterns

Module C5 — Chemicals of the Natural Environment

Module C6 — Chemical Synthesis

Module C7 — Further Chemistry

Exam Skills

Published by Coordination Group Publications Ltd.

Editors:
Amy Boutal, Ellen Bowness, Tom Cain, Katherine Craig, Sarah Hilton, Kate Houghton, Sharon Keeley, Andy Park, Kate Redmond, Rachel Selway, Ami Snelling, Jennifer Underwood, Julie Wakeling.

Contributors:
Michael Aicken, Mike Bossart, John Duffy, Richard Parsons, Andy Rankin, Philip Rushworth, Mike Thompson, Paul Warren, Sophie Watkins, Jim Wilson.

ISBN: 978 1 84762 009 5

With thanks to Jeremy Cooper, Barrie Crowther, Philip Dobson and Glenn Rogers for the proofreading.

With thanks to Laura Phillips for the copyright research.

With thanks to Science Photo Library for permission to reproduce the photograph used on page 10.

Groovy website: www.cgpbooks.co.uk

Printed by Elanders Hindson Ltd, Newcastle upon Tyne.
Jolly bits of clipart from CorelDRAW®

The Scientific Process

This section isn't about how to 'do' science — but it does show you the way most scientists work, and how scientists try to find decent explanations for things that happen. It's pretty important stuff.

Scientists Come Up with Hypotheses...

1) Scientists try to explain things. Everything.
2) Scientists start by observing or thinking about something they don't understand. It could be anything, e.g. what matter is made of, planets in the sky, a person suffering from an illness, ... anything.

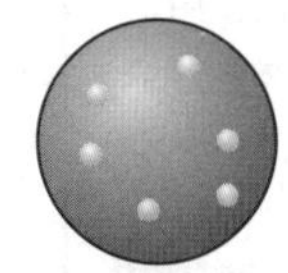

About 100 years ago, we thought atoms looked like this.

3) Then, using what they already know (plus a lot of creativity and insight), they work out an explanation (a hypothesis) that could explain what they've observed. Then they use their hypothesis to make a prediction that can be tested to provide further evidence to support the explanation. Whatever their explanation, it must stand up to the next stage of the scientific process — scrutiny from other scientists.

...Then Look for Evidence to Test Those Hypotheses

1) A hypothesis is just a theory — a belief. And believing something is true doesn't make it true — not even if you're a scientist.
2) So the next step is to try and find evidence to support the hypothesis. You might get evidence by doing controlled experiments in laboratories — a lab makes it easy to control variables so they're all kept constant (except the one you're investigating) — so it's easier to carry out a fair test.

Other Scientists Will Test the Hypotheses Too

1) Scientists report their findings to other scientists, e.g. by publishing their results in journals. Other scientists will carry out their own experiments and try to reproduce earlier results. And if all the experiments back up a hypothesis, then scientists start to have a lot of faith in it, and accept it as a theory.
2) However, if a scientist somewhere in the world gets results that don't fit with the hypothesis, either those results or the hypothesis must be wrong.

Then we thought they looked like this.

3) This process of testing a hypothesis to destruction is a vital part of the scientific process. Without the 'healthy scepticism' of scientists everywhere, we'd still believe the first theories that people came up with — like everything being made of air, water, fire or earth (or whatever).

If Evidence Supports a Hypothesis, It's Accepted — for Now

1) If pretty much every scientist in the world believes a hypothesis to be true because experiments back it up, then it usually goes in the textbooks for students to learn.
2) Our currently accepted theories are the ones that have survived this 'trial by evidence' — they've been tested many, many times over the years and survived (while the less good ones have been ditched).

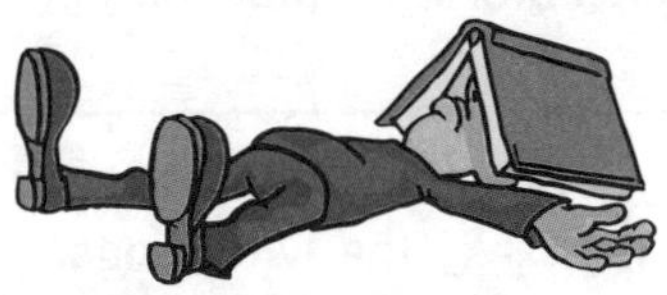

3) However... they never, never become hard and fast, totally indisputable fact. You can never know... it'd only take one odd, totally inexplicable result, and the hypothesising and testing would start all over again.
4) There isn't a scientific answer to everything yet — not by a long way.

Science is a "real-world" subject...

Science isn't just about explaining things that people are curious about — if scientists can explain something that happens in the world, then maybe they can predict what will happen in the future, or even control future events — to make life a bit better in some way, either for themselves or for other people.

The Scientific Process

You Need Reliable Data, Not Opinion, to Justify an Explanation

1) The only scientific way to test a hypothesis is to gather appropriate data.
2) And that data needs to be reliable — if it isn't, then it doesn't really help.

RELIABLE means that the data can be reproduced by others in independent experiments or studies.

If you'd made a new drug to treat malaria, say, you'd have to test it to make sure it worked. First of all, you'd try it on just a few people — 30, say. If it seemed to cure 25 of them, great, but you couldn't be confident that your data was reliable. (For example, just by chance those 25 people might have recovered anyway.) If lots of other scientists tested the drug on lots of other people and they all got similar results to you, then your result would be accepted.

Measurements aren't Always Accurate

You can't be sure that your measurements are accurate:

1) If you take a lot of measurements of the same thing, you won't always get the same result. For example, the pollution level in the air might differ depending on weather conditions, etc.
2) It could also be that the thing stays the same but your measurement of it changes — e.g. your measuring equipment isn't very accurate, or you're not very good at measuring.
3) The best way to get a good estimate of the result is to repeat the measurement lots of times and take an average. The values should all fall into a range that shows roughly where the real value is.
4) If you've got measurements that are obviously outside that range, then something might have gone wrong with those measurements.

A Correlation Doesn't Prove One Thing Causes Another

If there's a relationship between two things, then you can say there's a correlation between them, e.g.

- Sales of woolly hats increase when the weather is colder.
- The warmer water is, the more salt dissolves in it.

If you find a correlation between two things it's easy to think that one thing causes the other, but that's not always true — here's an example:

1) Primary school children with bigger feet tend to be better at maths. There's a correlation between the factor (big feet) and the outcome (better maths skills).
2) But it'd be crazy to say that having big feet causes you to be better at maths (and even weirder to say that being good at maths causes bigger feet...).
3) There's another (hidden) factor involved — their age. Older children are usually better at maths. They also usually have bigger feet. Age affects both their maths skills and the size of their feet.
4) If you really thought that there was a link between shoe size and ability at maths, you'd test this by comparing children of the same age — you have to control the other factors, so that the only factor that varies is foot size.

There might be a hidden factor influencing the results, or it could just be chance. Scientists don't usually accept a cause for something unless they can work out a plausible mechanism that links the two things.

All sheep die — Elvis died, so he must have been a sheep...

You read about correlations in the media all the time and reporters often make the mistake of thinking that if two things are correlated then one must cause the other. But that's often not the case at all.

Risk

By reading this page you are agreeing to the risk of a paper cut or severe drowsiness that could affect your ability to operate heavy machinery... Think carefully — the choice is yours.

Nothing is Completely Risk-Free

1) Everything that you do has a risk attached to it.
2) Scientists often try to identify risks — they show correlations (see previous page) between certain activities (risk factors) and negative outcomes.
3) Some risks seem pretty obvious, or we've known about them for a while, like the risk of getting heart disease if you're overweight, or of having a car accident when you're travelling in a car.
4) As new technology develops it can bring new risks, e.g. some scientists believe that using a mobile phone a lot may be harmful. There are lots of risks we don't know about yet.
5) You can estimate the size of a risk based on how many times something has happened in a big sample (e.g. 100 000 people) over a given period (say, a year). The more data you have to base your assessment on, the more accurate your estimate is likely to be.
6) There are two main parts to risk — the chances of something happening and how serious the consequences would be if it did. They usually add up to give an idea of how risky an activity is. So if something is very likely to happen and there are serious consequences it's high-risk.

People Make Their Own Decisions About Risk

1) Not all risks have the same consequences, e.g. if you chop veg with a sharp knife you risk cutting your finger, but if you go scuba-diving you risk death. You're much more likely to cut your finger during half an hour of chopping than to die during half an hour of scuba-diving. But most people are happier to accept a higher probability of an accident if the consequences are short-lived and fairly minor.
2) People are also more willing to accept risks if they get a significant benefit from the activity — e.g. car travel is quite risky, but the convenience of it means that people take the risk.
3) Freedom of choice plays quite a big part, too. People tend to be more willing to accept a risk if they're choosing to do something, rather than if they're having the risk imposed on them.
4) People's perception of risk (how risky they think something is) isn't always accurate. E.g. cycling on the roads is often high-risk, but many people are still happy to do it because it's a familiar activity. Air travel is actually pretty safe, but people perceive it as high-risk, so a lot of people are afraid of it.

We Have to Choose Acceptable Levels of Risk

1) People have to choose a level of risk that they find acceptable. This varies from person to person.
2) Governments and scientists often have to choose levels of risk in various situations on behalf of other people. They'll often be influenced by public opinion though.
3) Reducing risk can cost a lot, and it's not possible to reduce any risk to zero. The people responsible aim to keep the risks As Low As Reasonably Achievable (the ALARA principle).
4) Many people react to risk using the precautionary principle (although they might not realise it...):

THE PRECAUTIONARY PRINCIPLE

If you're not sure about the risks of something, but the results could be serious and irreversible, then it makes sense to try and avoid it, e.g. we're taking action now to try and prevent/slow down climate change, even though we don't know exactly what's going to happen.

Take a risk — turn the page...

So it all boils down to the probability of something happening and the consequences if it does. You can try to reduce a risk either by making it less likely to happen or by making the consequences less severe.

Science Has Limits

Science can give us amazing things — cures for diseases, space travel, heated toilet seats...
But science has its limitations — there are questions that it just can't answer.

Some Questions Are Unanswered by Science — So Far

1) We don't understand everything. And we never will. We'll find out more, for sure — as more explanations are suggested and more experiments are done. But there'll always be stuff we don't know.

   For example, today we don't know as much as we'd like about climate change.
   Is it definitely happening now? And what effects will it have on, say, our water supplies?

2) These are complicated questions. At the moment scientists don't all agree on the answers. But eventually, we probably will be able to answer these questions once and for all.
3) But by then there'll be loads of new questions to answer.

Other Questions Are Unanswerable by Science

1) Then there's the other type... questions that all the experiments in the world won't help us answer — the "Should we be doing this at all?" type questions. There are always two sides...
2) Think about new drugs which can be taken to boost your 'brain power'. These do exist — but does that mean we should allow people to take them?
3) Different people have different opinions. For example...

THE GAZETTE
BRAIN-BOOSTING DRUGS MAKE A MOCKERY OF EXAMS

THE POST
GENIUS PILLS TO BECOME THE NEW COFFEE

- Some people think they're good... Or at least no worse than taking vitamins or eating oily fish. They could let you keep thinking for longer, or improve your memory. It's thought that new drugs could allow people to think in ways that are beyond the powers of normal brains — in effect, to become geniuses...
- Other people say they're bad... taking them would give you an unfair advantage in exams, say. And perhaps people would be pressured into taking them so that they could work more effectively, and for longer hours.

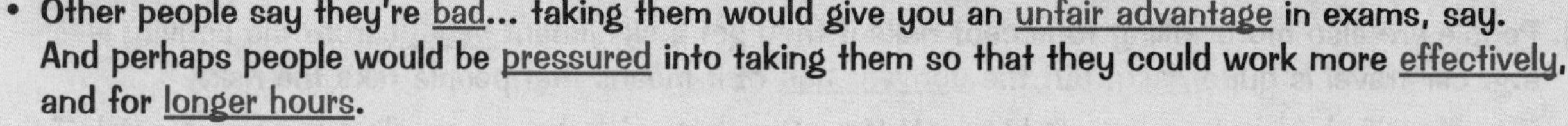

4) This question of whether something is morally or ethically right or wrong can't be answered by more experiments — there is no "right" or "wrong" answer.
5) The best we can do is get a consensus from society — a judgement that most people are more or less happy to live by. Science can provide more information to help people make this judgement, and the judgement might change over time. But in the end it's up to people and their conscience.

Loads of Other Factors Can Influence Decisions Too

Here are some other factors that can influence decisions about science, and the way science is used:

Economic factors:
- Companies very often won't pay for research unless there's likely to be a profit in it.
- Society can't always afford to do things scientists recommend without cutting back elsewhere (e.g. investing heavily in alternative energy sources).

Social factors:
- Decisions based on scientific evidence affect people — e.g. should fossil fuels be taxed more highly (to invest in alternative energy)? Should alcohol be banned (to prevent health problems)? Would the effect on people's lifestyles be acceptable...

Environmental factors:
- Pesticides and fertilisers help us produce more food cheaply — but some people are very concerned about the environmental problems they cause.

Science doesn't have all the answers...

Any scientific development will have benefits and costs for different groups of people. Some people argue that the right solution to an ethical problem is the one that leads to the best outcome for the majority of people involved. Others argue that some things are just wrong, and can never be justified.

The Atmosphere

We take air for granted — breathing it in and out and never giving it a second thought. But have a think about it — loads of things that we do every day are polluting and damaging our atmosphere.

The Atmosphere is a Mixture of Gases

1) The Earth is surrounded by an atmosphere — a mixture of gases.
2) The atmosphere is mostly made up of nitrogen, oxygen and argon:

Nitrogen 78%	Oxygen 21%	Argon 1%

3) The figures above are rounded slightly — the atmosphere also contains small amounts of carbon dioxide and various other gases, and varying amounts of water vapour.

This blue haze is the Earth's atmosphere.

Human Activity is Changing the Atmosphere

1) The concentrations of nitrogen, oxygen and argon in the atmosphere are pretty much constant.
2) But... human activity is adding small amounts of pollutants to the air. There are five pollutants you need to know about:

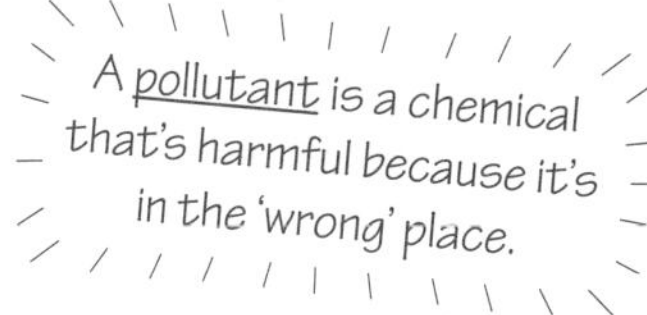

A pollutant is a chemical that's harmful because it's in the 'wrong' place.

- carbon dioxide (see p.7)
- carbon monoxide (see p.7)
- particles of carbon (see p.7)
- sulfur dioxide (see p.8)
- nitrogen oxides (see p.9)

3) These pollutants come from many different sources. You need to know about pollution from burning fuels — in power stations and vehicles. (See pages 7 to 9.)
4) Some pollutant gases are directly harmful to humans — they can cause disease or death in people who breathe in large enough quantities. E.g. vehicle exhausts contain pollutants which contribute to breathing problems like asthma.

5) Pollutants can also harm us indirectly, by damaging our environment. For example:

- Some pollutants cause acid rain, which pollutes rivers and lakes — killing the fish which people catch and eat.
- Other pollutants are thought to be leading to climate change, which could bring all sorts of problems — rising sea levels, disruption to farming, more hurricanes...

Most Fuels are Hydrocarbons

A compound is where different atoms are bonded together chemically.

Most of the fuels we burn in cars, trains, planes etc. are hydrocarbons.

1) A hydrocarbon is a compound containing hydrogen and carbon atoms only.
2) Fuels such as petrol, diesel fuel and fuel oil are mixtures of hydrocarbons.
3) Many power stations also burn hydrocarbons, e.g. natural gas.

Many other power stations burn coal. Coal isn't a hydrocarbon — it's just carbon (with a few impurities).

Polluting the atmosphere — smoking at parties...

The atmosphere's pretty important — without it, there'd be no fresh air to breathe. In some places though, the air isn't very 'fresh' at all. It's said that living in Mexico City does the same damage to your health as smoking 40 cigarettes a day — the city's built in a dip, so traffic fumes tend to hang over it.

Chemical Reactions

If you burn a fuel, a chemical reaction occurs — atoms from the fuel react with atoms from the air — and the atoms rearrange themselves to make other substances.

Chemical Reactions Happen When Atoms are Rearranged

Burning is a type of chemical reaction. Almost all chemical reactions involve atoms changing places.

1) When a hydrocarbon burns, the hydrogen atoms in the fuel combine with oxygen atoms from the air to make hydrogen oxide (otherwise known as water)...
2) ... and the carbon atoms in the fuel combine with oxygen from the air to make carbon dioxide.

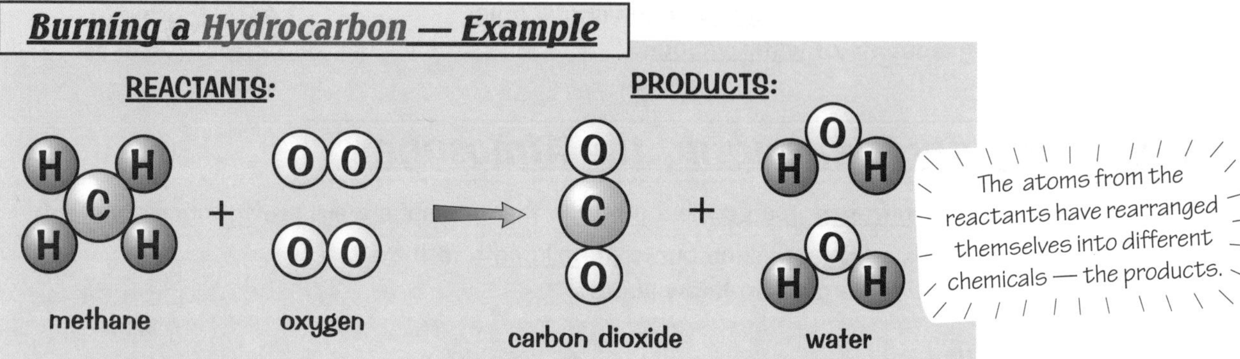

3) In the reaction shown above, the reactants are methane (a hydrocarbon) and oxygen...
4) ...and the products are carbon dioxide and water.
5) No atoms 'disappear' during the reaction — if you count up all the atoms in the reactants (1 carbon, 4 hydrogens and 4 oxygens), you'll find they're all still there in the products.
6) It's the same with any chemical reaction — the atoms get shuffled about, but they're all still there.
7) When coal burns, you mostly get carbon dioxide.

C + OO → OCO

carbon + oxygen → carbon dioxide

Whenever we burn fuels containing carbon (like coal, petrol, diesel, etc.) we add carbon dioxide to the atmosphere. But that's not all we add — fuels usually contain various impurities, which also combine with oxygen and make other pollutant gases.

Reactants and Products Often Have Very Different Properties

1) When atoms rearrange themselves in a reaction, the products that are formed have their own properties — which can be very different from the properties of the reactants.
2) For example, carbon is a black solid and oxygen is a colourless gas (at room temperature). When they react together, as coal burns, the product is carbon dioxide — which is a colourless gas, but with very different properties from oxygen (it's heavier for a start, and can be toxic).

The burning question is — do you know it all...

Chemical reactions are great — atoms rearrange themselves and you get totally different stuff from the stuff you started with. For example, sodium is a metal that will burn you if you touch it, and chlorine is a poisonous gas. But when they react together, you get sodium chloride — nice, edible table salt.

Air Pollution — Carbon

Coal, petrol, diesel, etc. are fossil fuels — they were formed from the remains of dead plants and animals.

Different Forms of Carbon Pollution Cause Different Problems

Fossil fuels are burnt to release energy. We burn fossil fuels to power vehicles and to produce electricity in power stations. The carbon-based products of burning fossil fuels often pollute the atmosphere.

1) All fossil fuels contain large amounts of the element carbon, so it's no surprise that these fuels produce a lot of pollutants that contain carbon.
2) If the fuel is burnt where there's lots of oxygen available, then nearly all the carbon ends up in carbon dioxide. This adds to the carbon dioxide that's found naturally in the atmosphere.
3) If there's not much oxygen available, such as in a car engine, then small amounts of carbon monoxide and small particles of carbon are produced as well.
4) Carbon dioxide, carbon monoxide and carbon particles are all pollutants.

Carbon Dioxide

1) Carbon dioxide has the formula CO_2, which means that a molecule of carbon dioxide has two oxygen atoms and one carbon atom.
2) In the same way, water's formula, H_2O, tells us that a molecule of water is made of two hydrogen atoms and one oxygen atom.
3) Carbon dioxide (CO_2), like any other atmospheric pollutant, will stay in the atmosphere causing problems until it's removed.
4) CO_2 can be removed from the atmosphere naturally. Plants use up CO_2 from the air when they photosynthesise. CO_2 also dissolves in rainwater and in seas, lakes and rivers.
5) Despite these ways of removing CO_2 from the atmosphere, CO_2 levels can still increase if human activity, e.g. burning fuels, adds extra CO_2 into the atmosphere.
6) Increased CO_2 levels increase the greenhouse effect, which is warming up the Earth. This may change the world's climate, possibly causing flooding due to the polar ice caps melting.

Carbon dioxide (CO_2)

C
O O

Water (H_2O)

O
H H

Carbon Monoxide

Unlike carbon dioxide, a carbon monoxide molecule has only one oxygen atom attached to a carbon atom. Carbon monoxide is produced if there's not enough oxygen available when fuels burn.

Carbon monoxide is poisonous — if a dodgy boiler in your home starts giving out carbon monoxide, it can make you drowsy and headachy, and can even sometimes kill.

Particulate Carbon

Often tiny particles of carbon are produced when fuels burn — this is called particulate carbon. If they escape into the atmosphere, which they often do, they just float around. Eventually they fall back to the ground and deposit themselves as the horrible black dust we call soot.

A lot of soot just falls onto buildings, making them look dirty, like this one.

Problems problems... there's always summat goin' wrong...

Energy companies are aware of the problems of burning fuels so they're investing money into developing cleaner, renewable sources of energy, which might not save the world but it'll help a lot. Hooray!

Air Pollution — Sulfur

Sulfur — it's dirty, nasty, pretty smelly, and you have to learn about it. No excuses.

Sulfur Pollution Comes from Impurities in Fuels

1) Many of our fuels are hydrocarbon-based, like petrol and natural gas. Some are just carbon-based, like coal. These fuels contain loads of impurities as they're extracted straight from the Earth's crust.
2) Many of these impurities are fairly harmless but some fuels contain traces of the element sulfur.
3) When the fuel burns, the sulfur burns too. When sulfur atoms burn, they combine with the oxygen in the air to produce the pollutant sulfur dioxide.
4) So when power stations and vehicle engines burn fossil fuels like coal and oil, small amounts of the pollutant sulfur dioxide are produced. This sulfur dioxide usually ends up in our atmosphere.

Look at all that gas — there's lots of sulfur dioxide in there, I bet.

Sulfur Pollution Causes Acid Rain

Sulfur dioxide has the formula SO_2. This tells you that a sulfur dioxide molecule is two oxygen atoms joined to one sulfur atom.

Sulfur dioxide (SO_2)

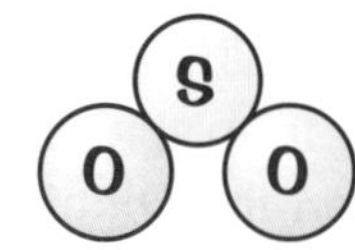

1) As with other pollutants, when sulfur dioxide gets into the atmosphere it will stay there until something gets rid of it.
2) The way sulfur dioxide usually leaves our atmosphere is in the form of acid rain.
3) When the sulfur dioxide emitted from vehicle engines and power stations reacts with the moisture in clouds, dilute sulfuric acid is formed.
4) Eventually, much of this acid will fall as acid rain, which is bad news for buildings, plants, animals and humans.
5) Acid rain causes lakes to become acidic, killing plants and animals. It also kills trees and damages buildings and statues made from some kinds of stone, e.g. limestone.

Acid rain — well at least it's more exciting than carbon...

Sulfur, for those of you who might think otherwise, is spelt with an 'f'. So don't go writing 'sulphur' any more because we're spelling it like the Americans these days. Unfortunately it wasn't my decision, but I do as I'm told and I'm afraid you should too. So, from here on, it's sulfur to us both.

Air Pollution — Nitrogen

Now for a new type of pollution. One that, surprisingly, is made from the nitrogen in the air itself.

Nitrogen Pollution Involves Nitrogen from the Air

Nitrogen pollution doesn't actually come from the fuel itself — it's formed from nitrogen in the air when the fuel is burnt.

1) Fossil fuels burn at such high temperatures that nearby atoms in the air react with each other.
2) Nitrogen in the air reacts with the oxygen in the air to produce small amounts of compounds known as nitrogen oxides — nitrogen monoxide and nitrogen dioxide.
3) This happens in car engines.
4) Nitrogen oxides are pollutants, and are usually spewed straight out into the atmosphere.

Nitrogen Oxides Are Nitrogen Monoxide and Nitrogen Dioxide

1) Nitrogen monoxide has the formula NO — it is made of one nitrogen and one oxygen atom.
2) Nitrogen dioxide has the formula NO_2 — it is made of two oxygen atoms joined to one nitrogen atom.
3) Nitrogen oxides (NO and NO_2) can be jointly referred to as NO_x.

Nitrogen monoxide (NO)

N O

Nitrogen dioxide (NO_2)

Here's how nitrogen oxides are formed:

NITROGEN MONOXIDE

Nitrogen monoxide forms when nitrogen and oxygen in the air are exposed to a very high temperature. This happens when fuels are burnt in places like car engines.

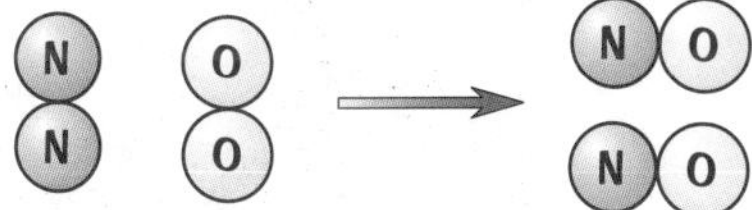

NITROGEN DIOXIDE

Once the nitrogen monoxide is in the air, it will go on to react with more oxygen in the air to produce nitrogen dioxide.

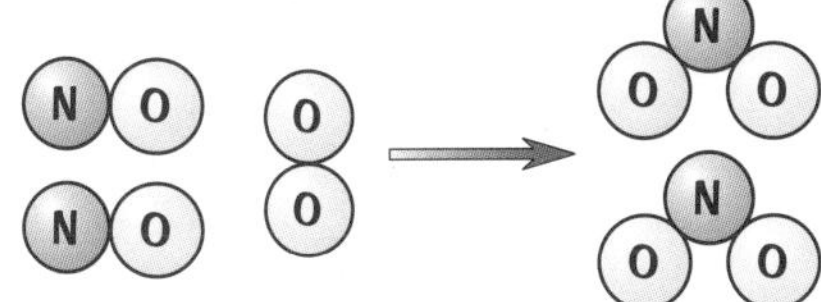

As pollutants, nitrogen oxides are very similar to sulfur dioxide. When they're formed they usually end up in the atmosphere — which is where they stay until they react with moisture in clouds. This produces a dilute nitric acid which eventually falls to the Earth as acid rain.

Acid rain makes statues talk too.

I'll pollute you if you don't learn this...

Oh good — another type of pollution to learn about. Well, in my opinion, the only answer is to go back to the old days of farming a bit of land in a self-sufficient manner. No more electricity, no more industry, bliss. Although there'd be no more TVs, DVD players, X-ray machines, computers. Hmm, in that case...

Reducing Pollution

There are plenty of things we can be doing to try and reduce air pollution.

We Can Reduce Pollution from Power Stations...

The simplest way to reduce the amount of all types of pollution from power stations is to use less electricity. The less electricity we use, the less fuel needs to be burnt in power stations and so the less pollution is created. The problem is that we've all got used to consuming lots of electricity — it would be hard to convince people to do without TVs, washing machines, computers, etc.

But even without cutting right back on our electricity usage, there are ways to reduce pollution:

1) Much of the sulfur can be taken out of the natural gas and fuel oil that power stations use. This means that little sulfur dioxide is produced when it burns.
2) When coal is burnt in power stations, most of the sulfur dioxide and the particulates (carbon particles and ash) can be removed before they can get into the atmosphere.
3) The only way to reduce CO_2 emissions is to reduce the amount of fossil fuels burnt, either by using less electricity, or by finding alternative energy sources.

...And from Our Cars

1) Motor vehicles now have more efficient engines, which burn less fuel and so create less pollution.
2) Low-sulfur fuel for cars is now available, which means less sulfur dioxide is emitted from the exhaust.
3) Many cars are now fitted with catalytic converters. These convert harmful nitrogen monoxide into harmless nitrogen and oxygen. They also convert the very toxic gas carbon monoxide into the less harmful gas carbon dioxide.

This is a catalytic converter.

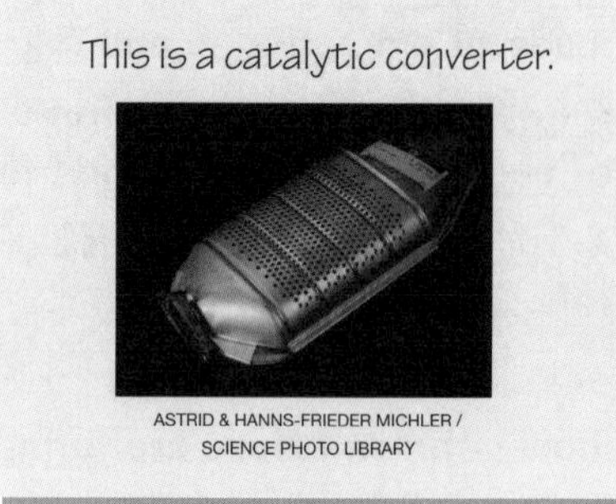

ASTRID & HANNS-FRIEDER MICHLER / SCIENCE PHOTO LIBRARY

Cars more than 3 years old must pass an MOT test once a year to prove they're safe and not too polluting.

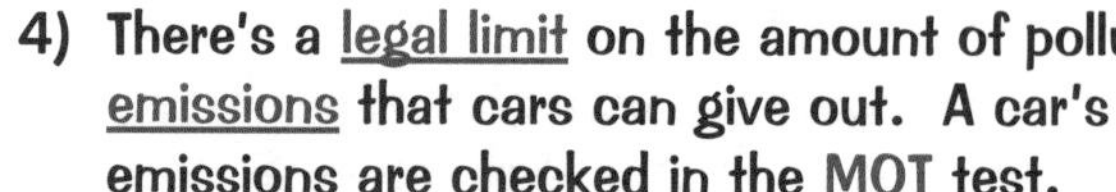

4) There's a legal limit on the amount of polluting emissions that cars can give out. A car's emissions are checked in the MOT test.
5) We could also reduce pollution by using our cars less. If everyone used public transport instead of individual cars then less petrol would be burnt overall.

Laws and Regulations Help Reduce Pollution

Certain laws and regulations exist to make individuals and organisations responsible for the pollution they produce.

Any process that is likely to add pollutants to the atmosphere, including scientific research, must comply with laws and regulations that limit the amount of polluting chemicals that can be released into the atmosphere (and elsewhere).

My brother is a major source of air pollution...

So... it doesn't sound like reducing pollution is going to be easy. We need to reduce the demand for electricity by not using so much of it, and then we won't need so many power stations. Start small by remembering to switch off lights, and don't leave stuff on stand-by — that's just wasting electricity.

Interpreting Pollution Data

Here's a page on the reliability of scientific research and why you shouldn't believe everything you read in the papers.

Claims About Air Pollution Can be Hard to Justify

Lots of studies have linked air pollution to various health problems.
When these claims are made though, we shouldn't just take the results at face value.

Say some scientists have claimed that the number of asthma attacks in children is linked to the level of a particular pollutant in the air.

Don't be too convinced about the results if...

...It's Just One Small Study That Hasn't been Repeated

1) The scientists would have to provide details of the experiments they did, the size of their sample and the results they got. The bigger the sample used in the study, the more confident you can be about the results.
2) That's not enough though. Other scientists would need to repeat the experiments and get the same results before the claims could be accepted by the scientific community.
3) A study's results are only really trustworthy if other scientists can replicate them — that way you know that the first study didn't just come up with freak results.

...It Doesn't Take Other Factors into Account

Any outcome (e.g. an asthma attack) could be caused by factors other than the one you're investigating (e.g. air pollution) — air pollution isn't the only cause of asthma attacks.

Pollen — it has a lot to answer for.

1) It's thought that asthma attacks can be triggered by factors like high pollen levels, infections (e.g. colds and flu) and emotional stress.
2) Any study into the effect of air pollution levels on the number of asthma attacks needs to account for these factors in some way, e.g.
   - The study should be carried out at a time of year when the pollen levels are approximately the same in high pollution and low pollution areas — so probably not spring or summer.
   - If any of the subjects get over-stressed or ill in a way that might affect the results, they should be excluded from the study.

...The Conclusion Goes Beyond the Data

1) A correlation between two things doesn't necessarily mean that one caused the other (take a look back at page 2 for more about correlation and cause).
   So a correlation between air pollution levels and the number of asthma attacks doesn't necessarily mean that the air pollution is causing the attacks.
2) Also remember that if one thing does cause something else, it doesn't mean that it will definitely make it happen. So if the study did show that air pollution causes asthma attacks, all that means is that your risk of having an asthma attack is increased in a high pollution area.

There's a correlation between madness and chemistry knowledge...

Don't just believe what you read — ask questions like 'but how reliable was the experiment?', 'but could anything else have caused those results?' and 'but does that correlation show causation?'. Have an enquiring mind. This is how science works — scientists look for evidence for or against theories.

Sustainable Development

Sustainable development is a hot topic at the moment — it's relevant to subjects like Geography too.

More People Means More Pressure on the Environment

1) The world's population is increasing at an amazing rate.
2) It has doubled in the last 40 years and now totals 6.5 billion. It shows no sign of slowing down.
3) The more of us there are, the more food must be produced, the more fuel is burnt and the more land is used up. This all puts a lot of pressure on the environment.

Scientific Advances May Have Helped Cause the Problems...

Science-based technology has improved the quality of life for most of us, but is also unfortunately contributing to environmental problems. Computers are a good example of this:

1) Computers have made workplaces more efficient, and the internet has improved global communication, which is great for both individuals and businesses. Computers are still getting better all the time.
2) The trouble is that computers are full of poisonous substances. Computers don't last for ever and these poisonous substances cause a problem when it comes to computer disposal.
3) Many old computers are sent to poorer countries to be dismantled. There they're often dumped in landfills. This poisons the land for the people, animals and plants that live there.

Computers are great, as are many of the scientific advances that make our lives easier. However, the cost to the environment and ultimately to humans is massively important. Benefits need to be weighed against costs when deciding whether to use new technologies.

...But Other Advances Could Help Us Live More Sustainably

Often science comes up with solutions to the problems it causes. In particular, science has helped us find ways of using our natural resources to meet our needs without messing things up so that future generations can't meet theirs. This is known as sustainable development. For example:

1) Scientific advances have allowed us to develop buildings that use less energy. For example, modern buildings have insulated windows and doors which mean less energy is used to heat them.
2) Using less energy means less fossil fuels are burnt in the building or at power stations and so less pollution reaches the atmosphere.
3) Buildings can also use renewable resources as energy sources, e.g. solar energy can be used to heat water and generate electricity. Wind turbines on the roofs of houses can be used to generate electricity.

These days, there are laws and regulations governing scientific research. In many areas of science, scientists have to show that what they're developing will not go on to have an unacceptable impact on the environment. This particularly applies to pollutants being released into the atmosphere.

Try wind power — it can blow your mind...

It's kind of obvious that scientific development has led to environmental problems. I mean, back when there was little industry but lots of farming, fishing and stuff, there was little pollution. But since then, lots of scientific advances have caused problems — from pollution to cutting down rainforests.

Revision Summary for Module C1

Congratulations! You've made it through C1. You should now be an expert on fossil fuels, carbon, sulfur, air pollution and the general problems we're causing to the environment and our own health. You should also have picked up a little bit on atoms and what happens to them in chemical reactions. Anyway, maybe it's time you found out how much you do know. A good try at the questions below should give you some idea. Here we go...

1) What three main gases is the atmosphere made of and in what proportions are they found?
2) Name one result of pollution which causes direct harm to humans, and one which causes indirect harm.
3) What is a hydrocarbon?
4) Name the main elements that make up coal.
5) When a hydrocarbon fuel burns, with what substance in the air do the hydrogen and carbon atoms combine?
6) When hydrocarbons burn in plenty of oxygen, the atoms involved rearrange themselves into carbon dioxide and what else?
7) Atoms can disappear completely in some chemical reactions — true or false?
8) Are the properties of reactants and products always the same or can they be different?
9) What removes carbon dioxide from the atmosphere?
10) What atoms make up carbon monoxide? Under what conditions is carbon monoxide produced?
11) What are particles of carbon otherwise known as and what kind of pollution do they cause?
12) Describe briefly how the pollutant sulfur dioxide is produced. Coal / Petrol
13) How does sulfur dioxide leave the atmosphere?
14) What effects does acid rain have on the environment?
15) Describe how nitrogen dioxide is produced when fossil fuels are burnt.
16) Give the formulas of nitrogen monoxide and nitrogen dioxide.
17) What effect do nitrogen oxides have on the environment?
18) What do catalytic converters do?
19) Name two things that everyone could do in order to reduce carbon dioxide pollution.
20) How do MOT tests help combat air pollution?
21)* Why should you question the results of an experiment if the experiment was only done once?
22) If both air pollution and cancer increase, does that mean air pollution causes cancer?
23) Why is it a problem that the human population is rapidly increasing?
24) Give an example of a scientific advancement that contributes to environmental problems.
25) What is sustainable development?
26) Give an example of a renewable resource that could be used to provide energy for our homes.

* Answers on page 92.

Natural and Synthetic Materials

This section is all about materials, their properties and the best ones to use to make different products.

All Materials are Made Up of Chemicals

Absolutely everything is made up of chemicals. The materials we use are either made of individual chemicals or mixtures of chemicals.

Chemicals are made up of atoms or groups of atoms bonded together.

1) Iron is a chemical element — it's made up of iron atoms.

2) Water is a chemical. It's made up of lots of water molecules. A molecule is a group of atoms bonded together. Water molecules contain 2 chemical elements — hydrogen and oxygen.

Some materials are mixtures of chemicals. A mixture contains different substances that are not chemically bonded together. For example, rock salt is a mixture of two compounds — salt and sand.

Some of These Materials Occur Naturally...

A lot of the materials that we use are made by other living things, not by humans:

MATERIALS FROM PLANTS

1) Wood and paper are both made from trees.
2) Cotton comes from the cotton plant.

MATERIALS FROM ANIMALS

1) Wool comes from sheep.
2) Silk is made by the silkworm larva.
3) Leather comes from cows.

...Others are Synthetic — Made by Humans

We often use man-made (synthetic) materials instead of natural materials, e.g.

1) All rubber used to come from the sap of the rubber tree. We still get a lot of rubber this way (e.g. for car tyres), but you can also make rubber in a factory. The advantage of this is that you can control its properties, making it suitable for different purposes, e.g. wetsuits. However, it is cheaper to use rubber from the sap of the rubber tree.

2) A lot of clothes are made of man-made fabrics like nylon or polyester. These materials are often a lot cheaper and have more uses than natural fabrics like wool and silk — e.g. you can make fabrics that are super-stretchy, or sparkly.
3) Most paints are mixtures of man-made chemicals. The pigment (the colouring) and the stuff that holds it all together are designed to be tough and to stop the colour fading.

So silk comes out of a worm's bottom then...

OK, so now you should know what a material is (and that it doesn't just mean fabric). You should also have a pretty good idea of the different kinds of materials around, and where they come from. Not bad for the first page of the section. Time to cover the page and scribble down what you can remember.

Materials and Properties

Not all materials are the same, as you'll find out if you try to make a hat out of spaghetti hoops...

Different Materials Have Different Properties

MELTING POINT

Most materials that are pure chemicals have a unique melting point. This is the temperature where the solid material turns to liquid. E.g. the melting point of water is 0 degrees Celsius (0 °C).

STRENGTH

Strength is how good a material is at resisting a force. You can judge how strong it is by how much force is needed to either break it or permanently change its shape (deform it). There are two types of strength you need to know about:

1) TENSILE (OR TENSION) STRENGTH — how much a material can resist a pulling force. Things like ropes and cables need a high tensile strength, or they'd snap.
2) COMPRESSIVE STRENGTH — how much a material can resist a pushing force. Building materials like bricks need good compressive strength, or they'd be squashed by the weight of the bricks above them.

Some things, like cross beams in roofs, need to be made of materials with good compressive and tensile strength — they get both pushed and pulled.

STIFFNESS

A stiff material is good at not bending when a force is applied to it. This isn't the same as strength — a bendy material can still be strong if a big force doesn't permanently deform it.

1) Materials like steel are very difficult to bend — they're very stiff.
2) Some kinds of rubber are very strong but they bend and stretch very easily — they're not stiff.

HARDNESS

The hardness of a material is how difficult it is to cut into.

1) The hardest material found in nature is diamond.
2) The only material that can cut a diamond is another diamond.
3) Diamonds can cut most other materials — many industrial drills have diamond tips.

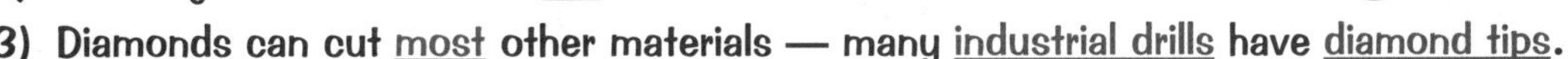

DENSITY

Density is a material's mass per unit volume (e.g. g/cm^3). Don't confuse density with mass or weight.

1) Air is not very dense. You'd need a huge volume of it to make up 1 kg in mass.
2) Gold is very dense. A small volume of gold would make up 1 kg in mass.
3) Objects that are less dense than water will float (like ice). Objects that are more dense than water will sink.

Substance	Density g/cm^3
gold	19.3
iron	7.9
PVC	1.3
water	1.0
ice	0.97
air	0.001

'Cos diamonds are an industrial drill's best friend...

Learn everything on this page — I make that five things. And make sure you're clear about what density is and why it's NOT the same as mass. Measure out 1 kg of loose change (or rocks) and 1 kg of breakfast cereal and compare the different volumes — same mass, but very different densities...

Making Measurements

When you measure the properties of a material you can only accept the results as reliable if other scientists elsewhere can get similar results — your experiment has to be repeatable.

Most Measurements Involve a Degree of Uncertainty

When you're measuring something there are loads of reasons why your results might not be accurate:

1) There might be a fault in your equipment.
2) Wrong results can be because of human error — inaccurate measuring, reading or recording of the results.
3) Your samples and techniques have to be the same every time. If lots of scientists all test the strength of iron they might all get different results because iron's strength depends on things like how it's made. The scientists need to work on an identical sample using identical techniques.

Measurements Will Always Vary to Some Extent

If you want an accurate result you've got to take measurements several times. This allows for faults in the equipment, human error, etc. You won't get exactly the same measurement each time, but that's normal.

This table shows the results of an experiment to measure the density of gold — the measurement was repeated 10 times:

1) The results of tests C and E are so different from the others that something must have gone wrong. Results like these that are an abnormal distance from the rest of the data are called outliers, and you can often just ignore them.
2) Working out what to ignore can be tricky — you need to be able to show that there's a 'real difference' between the measurements before you can ignore them. Plotting the results on a graph can help with this — you can draw a 'line of best fit' and ignore the results that aren't near it.
3) You can't always ignore outliers — if you're expecting the data to vary a lot, e.g. if you're measuring the height of children aged 1 to 10, then you wouldn't ignore a very low or high result.
4) The table suggests that the true value of the density of gold is in the range of 19.1 – 19.5 g/cm^3, where most of the measurements lie.
5) A mean (average) gives you the best estimate of the true value. Take a mean of the remaining results (add them together then divide by the number of good results (8)). This works out at 19.3 g/cm^3.

Test	Density of gold g/cm^3
A	19.3
B	19.4
C	12.8
D	19.1
E	30.1
F	19.2
G	19.5
H	19.2
I	19.2
J	19.5

Experiments Need to be Carefully Designed

1) Measuring a mean and ignoring outliers isn't enough. You also need to make sure that your experiment is a fair test.
2) The best way to make it a fair test is to vary only one factor in your experiment, and to only measure one thing at a time.
3) So if you're measuring densities of different materials, the only thing that you should vary each time is the material — that's the factor that you change. The volume of material that you test, the temperature and equipment that you use etc. must all be exactly the same each time.
4) Each time you repeat the test the other factors must be exactly the same — they must be controlled.

I'm studying for a degree of uncertainty right now...

This page is mostly about making sure that measurements are accurate, but it applies to any kind of scientific test really. If you don't follow these rules, then you can't really conclude anything from your results. These rules have to be followed by everyone, even the top scientists — so learn them now.

Materials, Properties and Uses

Every material has a different set of properties, which makes it perfect for some jobs, and totally useless for others. That probably explains why chocolate teapots have never really caught on...

The Possible Uses for a Material Depend on Its Properties

When you're choosing a material to use in a product, you need to think about its properties, e.g.

PLASTICS

- Can be fairly hard, strong and stiff
- Some are fairly low density (good for lightweight goods)
- Some are mouldable (easily made into things)

E.g. cases for televisions, computers and kettles

RUBBER

- Strong but soft and flexible
- Mouldable

E.g. rubber car tyres

NYLON FIBRES

- Soft and flexible
- Good tensile strength

E.g. ropes and clothing fabric

A Product's Properties Depend on the Materials It's Made From

Some products can be made from a variety of materials. How good the product is and how long it lasts depends on the properties of the materials it's made from.

1) Gramophone records 100 years ago were made of a mixture of natural materials like paper, slate and wax. There aren't many of these records left because they broke very easily — they weren't strong.
2) More modern records are made of polyvinyl chloride (PVC) or 'vinyl'. This material is strong and flexible so it's less likely to break. DJs sometimes still use vinyl records in clubs.
3) Most people these days own compact discs (CDs). These are made of a very tough, flexible plastic called polycarbonate. It's quite strong and hard (it's used in bulletproof glass) and should last even longer than PVC — but we'll have to wait a while to find out...

You'll Need to Assess the Suitability of Different Materials

You need to be able to look at the properties of a material and work out what sort of purposes it might be suitable for, e.g.

1) Cooking utensils must be made from something with a high melting point that's non-toxic.
2) Material to make a toy car must be non-toxic and should be strong, stiff and low density — e.g. some kinds of plastic.
3) Clothing fabric mustn't be stiff, but needs a good tensile strength (so it can be made into fibres) and high flame-resistance, especially if it's for nightwear or children's clothes.

It's not rocket science — that's 'cos it's materials science...

It's all fairly straightforward stuff on this page — just be prepared to look at the properties of 'mystery' materials in the exam, and work out which material would be most suitable for different jobs. Don't think that any of the materials are totally useless — their properties will probably be ideal for something.

Chemical Synthesis and Polymerisation

Crude oil is formed from the buried remains of plants and animals — it's a fossil fuel. Over millions of years, with high temperature and pressure, the remains turn to crude oil, which can be drilled up.

Crude Oil is a Mixture of Lots of Substances

Crude oil is a mixture of hydrocarbons — molecules which are made of chains of carbon and hydrogen atoms only (see p.5 for more info). These chains are of varying lengths.

Crude Oil is Used to Make Loads of Synthetic Substances

1) Most of the hydrocarbons in crude oil are refined by the petrochemical industry to produce fuels and lubricants.
2) Only a very small amount of hydrocarbons from crude oil are chemically modified to make new compounds for use in things like plastics, medicines, fertilisers and even food.

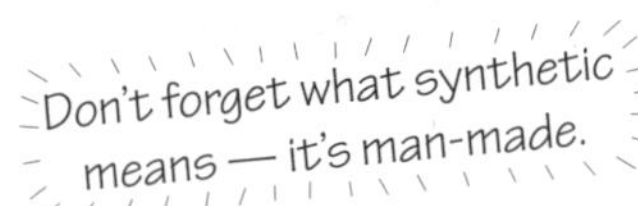

Don't forget what synthetic
means — it's man-made.

Polymerisation Means Loads of Small Molecules Link Together

Polymers are among the most important man-made materials — you can make loads of useful stuff from them — well, as long as you want a plastic-type thing, anyway. And I often do.

1) Plastics are formed when lots of small molecules join together to give a polymer.
2) They're usually carbon-based.
3) Under high pressure many small molecules 'join hands' (polymerise) to form long chains called polymers.

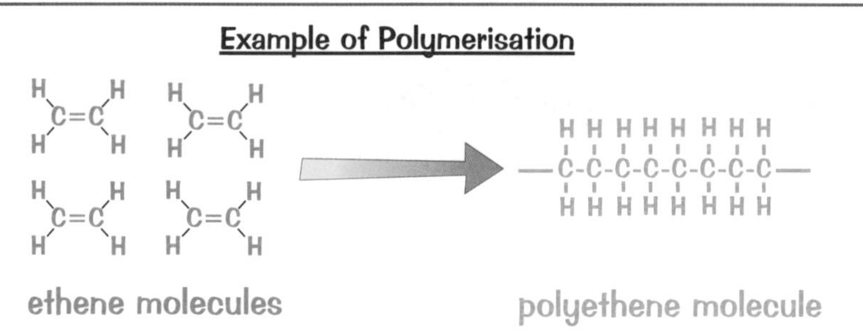

Different Polymers Have Different Properties

Different polymers have different physical properties — some are stronger, some are stretchier, some are more easily moulded, and so on. These different physical properties make them suited for different uses.

- Strong, rigid polymers such as high-density polyethene are used to make plastic milk bottles.
- Light, stretchable polymers such as low-density polyethene are used for plastic bags and squeezy bottles. Low-density polyethene has a low melting point, so it's no good for anything that'll get very hot.

- PVC is strong and durable, and it can be made either rigid or stretchy. The rigid kind is used to make window frames and piping. The stretchy kind is used to make synthetic leather.
- Polystyrene foam is used in packaging to protect breakable things, and it's used to make disposable coffee cups (the trapped air in the foam makes it a brilliant thermal insulator).
- Heat-resistant polymers such as melamine resin and polypropene are used to make plastic kettles.

Revision — it's all about stringing lots of facts together...

If you're making a product, you need to pick your plastic carefully. It's no good making a kettle from plastic that melts at 50 °C — you'll end up with a messy kitchen, a burnt hand and no cuppa. You'd also have a bit of difficulty trying to wear clothes made of brittle, unbendy plastic.

Structures and Properties of Polymers

You need to know how the properties of a polymer are affected by the way it's made.

A Polymer's Properties Decide its Uses

Its Properties Depend on How the Molecules are Arranged...

A polymer's properties don't just depend on the chemicals it's made from.
The way the polymer chains are arranged has a lot to do with them too:

If the polymer chains are packed close together, the material will have a high density.
If the polymer chains are spread out, the material will have a low density.

...And How They're Held Together

The forces between the different chains of the polymer hold it together as a solid mass.

Weak Forces:
Chains held together by weak forces are free to slide over each other. This means the plastic can be stretched easily, and will have a low melting point.

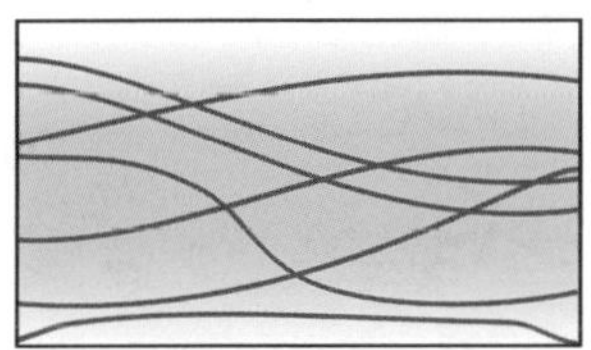

Strong Forces:
Plastics with stronger bonds between the polymer chains have higher melting points and can't be easily stretched, as the crosslinks hold the chains firmly together. Crosslinks are chemical bonds between the polymer chains (see below).

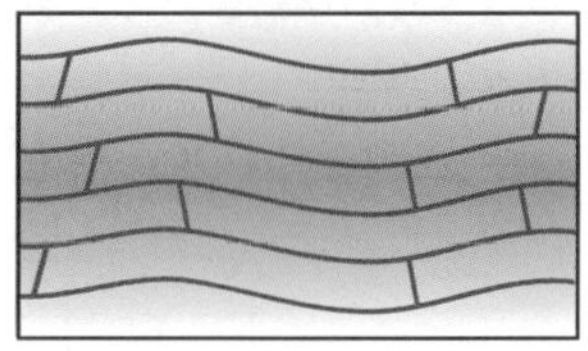

So, the stronger the bonds between the polymer chains, the more energy is needed to break them apart, and the higher the melting point.

Polymers Can be Modified to Give Them Different Properties

You can chemically modify polymers to change their properties.

1) Polymers can be modified to increase their chain length. Polymers with short chains are easy to shape and have lower melting points. Longer chain polymers are stiffer and have higher melting points.
2) Polymers can be made stronger by adding cross-linking agents. These agents chemically bond the chains together, making the polymer stiffer, stronger and more heat-resistant.

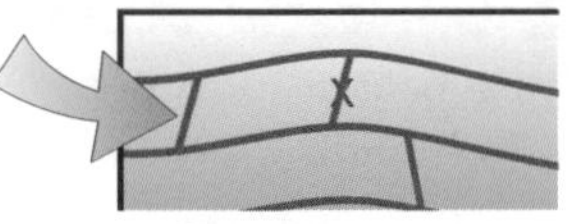

3) Plasticisers can be added to a polymer to make it softer and easier to shape. Plasticisers work by getting in between the polymer chains and reducing the forces between them.

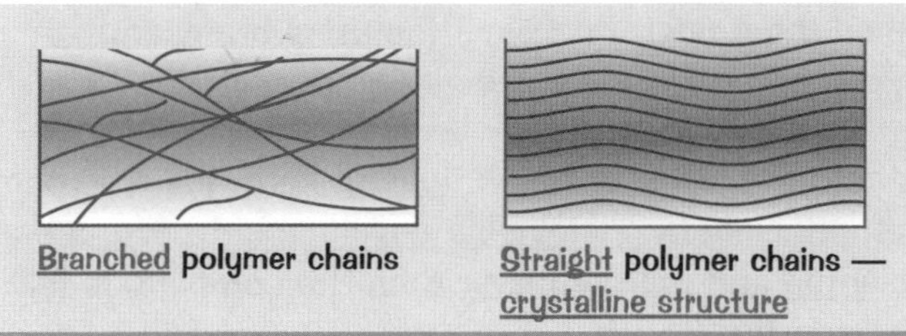

4) The polymer can be made more crystalline. A crystalline polymer has straight chains with no branches so the chains can fit close together. Crystalline polymers have higher density, are stronger and have a higher melting point.

Choose your polymers wisely...

The molecules that make up a plastic affect the properties of the plastic, which also affects what the plastic can be used for. I know there's a lot of diagrams of lines on this page, but the structure of polymers is really important — it affects their properties, and it might even come up in the exam.

Life Cycle Assessments

If a company wants to manufacture a new product, they carry out a life cycle assessment (LCA). This looks at every stage of the product's life to assess the impact it would have on the environment.

Life Cycle Assessments Show Total Environmental Costs

A life cycle assessment (LCA) looks at each stage of the life cycle of the product from the raw materials to when it's disposed of, and works out the potential environmental impact:

1) Extracting and refining raw materials
   E.g. metals may have to be mined from the ground, then extracted from their ore. Polymers come from crude oil, which has to be drilled and refined. All these things use energy, which usually means burning fossil fuels. Sometimes there's a limited supply of the raw material, too — e.g. oil supplies may run out soon.

2) Manufacturing the product
   This often uses a lot of energy and may cause pollution and use other resources — e.g. making a new car uses about 9000 litres of water.

3) Using the product
   Just using the finished product can also damage the environment — e.g. an electrical product like a TV uses electricity made from burning fossil fuels.
4) Disposing of the product
   When people have finished with the product, it has to be disposed of. It might be incinerated (burnt) which might cause air pollution, or it might go into a landfill site, or be recycled. All these options have an environmental impact.

Life Cycle Assessments are Helpful for Making Decisions

An LCA helps you work out the best materials and manufacturing process for your product. If the environmental impact or the cost is too high, you can choose another material or manufacturing process. If it's quite low, you might decide to use the materials for other products too.

The LCA can tell you if it's possible to make a product, and what the environmental impact will be. It can't tell you if you SHOULD make the product though — you have to consider other things when you're making your decision:

1) Making the product will benefit some people, like employees and customers — people need jobs, and some products are essential to their customers, e.g. syringes for diabetes sufferers.
2) Some people may be badly affected, e.g. by land pollution when the product is disposed of, or pollution from the mine where the raw material comes from.
3) Many countries have laws which limit how much impact a company can have on the environment. Poorer countries may really need the money that manufacturing brings in. Their governments are under pressure to be less strict about environmental concerns.

Companies can use the information from an LCA to set up a process which doesn't harm the environment so much that future generations suffer. This is a key principle of sustainable development. E.g. a paper company only using wood from forests that are replanted and regrow faster than the company is felling them, and taking steps to protect the wildlife that lives in the forests.

Need exercise? Go life-cycling then...

The environmental cost of a product can vary a lot. If it's something that goes out of date quite quickly, like a computer, then it will have a short life cycle. However, high quality expensive furniture might be kept for a lifetime, and its long life cycle mean it has less impact on the environment.

Revision Summary for Module C2

This section has gone from silk and rubber to crude oil and then onwards to environmental issues. That's an awful lot to take in in one section. Whether you find this topic easy or hard, interesting or dull, you've simply got to learn it all before the exam. Try these questions and see how much you really know.

1) Give one example of a chemical element.
2) Name a material we use that comes from:
   a) plants, b) animals.
3) Why do we use both natural and synthetic rubber?
4) What's the difference between compressive and tensile strength?
5) What's the hardest material found in nature?
6) Give a definition of density.
7) You're measuring the properties of a material — give three reasons why your results might not be accurate.
8) Why should you always repeat your measurements?
9)* You do an experiment to measure the density of a mystery material. You do the measurements nine times, and your results (in g/cm^3) are as follows: 8.1, 8.3, 8.1, 8.0, 14.2, 8.3, 8.4, 8.0, 8.2.
   a) Which result would you discard from your data?
   b) What is the mean of your measurements of the density of the mystery material?
10) How can you ensure that an experiment is a fair test?
11) Name two properties of each of the following materials that make them useful in manufacturing:
   a) plastic, b) rubber, c) nylon.
12)* Name three properties that you'd look for when choosing a material to make a child's dinner bowl.
13) What does crude oil contain?
14) Briefly describe what happens during polymerisation.
15) Give an example of a product where a polymer has replaced a natural material.
16) How does the arrangement of polymer chains affect the density of a material?
17) A polymer is easily stretched and has a low melting point. What can you say about the arrangement of its molecule chains and the forces holding them together?
18) What would you add to a polymer to make it stiffer and stronger?
19) How do plasticisers work?
20) What are the properties of a crystalline polymer?
21) Name the four main stages considered in a life cycle assessment.
22) Explain why a Life Cycle Assessment can't tell you whether you should make a product or not.
23) Give an example of sustainable development.

* Answers on page 92.

Recycling Elements

Thought recycling was a new idea developed by hippies in the 1970s? It turns out that clever old Mother Nature's been recycling stuff for millions of years. And she never wore tie-dye either.

Elements are Constantly Being Recycled

There isn't a never-ending supply of the elements that living things need — luckily elements are recycled:

1) As plants grow they take in elements like oxygen, nitrogen and carbon through their leaves and roots.
2) When the plants die and decompose, most of these elements are returned to the soil. Others go into the air as gases like methane.
3) Some of the elements in plants become part of animals when the plants are eaten. These elements are also returned to the environment when the animals poo or when they die and decompose.
4) Dead animal and plant matter (and animal waste) is broken down by microbes. They convert it into compounds that are taken up by other plants and the whole process starts again.

The Nitrogen Cycle is a Good Example of Recycling an Element

The constant cycling of nitrogen through the atmosphere, soil and organisms is called the nitrogen cycle.

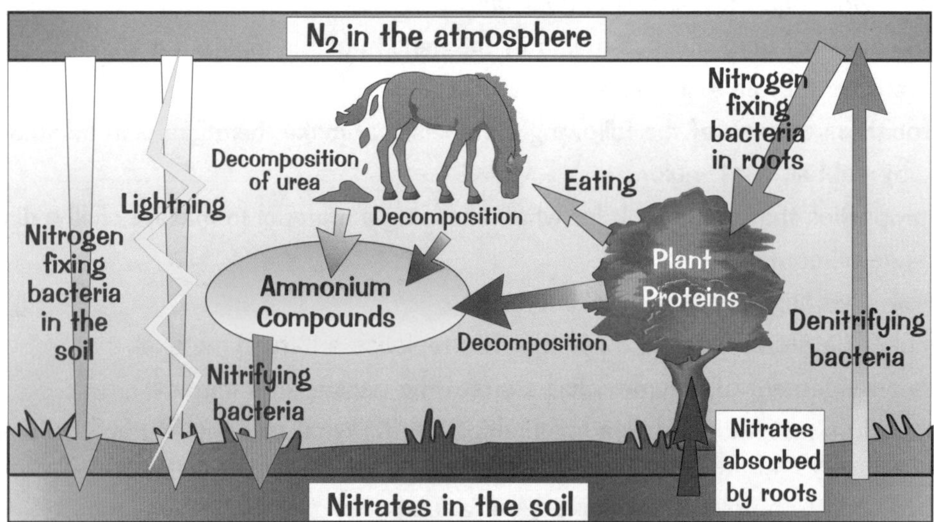

1) Plants absorb nitrogen in the form of nitrates from the soil. Nitrogen is needed by plants and animals so that they can make proteins (see p.25).
2) Animals have to eat plants (or other animals) to get their nitrogen.
3) Any organic waste, i.e. rotting plants, dead animals and animal poo, is broken down by microbes called decomposers into ammonium compounds.
4) Nitrifying bacteria turn the ammonium compounds produced by the microbes into useful nitrates.
5) These nitrates can then be absorbed by the roots of green plants once again.
6) A couple of extra weird bits worth mentioning — nitrogen-fixing bacteria live in the soil and the roots of some plants and can cleverly make nitrates directly from nitrogen in the air. Energy from lightning can also make nitrogen and oxygen in the air react to give nitrates in the soil. But denitrifying bacteria do the opposite and break down nitrates in the soil to give nitrogen in the air again.

It's the cyyyycle of liiiiife...

I guess it's pretty obvious really — there's only so much stuff on Earth and so it has to be recycled. Make sure you understand the basic principles at the top of the page and the details of the nitrogen cycle. By the way, there are about a million different ways of drawing the nitrogen cycle, so don't be put off if there's one in the exam that looks a bit different from this one — the basic ideas are the same.

Organic and Intensive Farming

Thought farming was all much of a muchness? Well think again, because there are two distinctly different ways of producing food — organic farming and intensive farming.

Harvesting Crops Removes Elements from the Soil

All crops are eventually harvested and removed from the field so that they can be sold and eaten.

1) The removal of the crops means that some of the elements that the plants used to grow are taken out of the field for good, instead of being returned to the soil when the plants die and decay.
2) Elements that are lost include nitrogen, phosphorus and potassium.
3) These elements need to be replaced or the fertility of the soil will decrease and the next crop of plants won't grow properly.
4) The way that farmers replace the lost elements depends on whether they farm organically or intensively.

Different Farming Methods Replace Lost Elements Differently

1) Organic Farming Relies on the Natural Recycling of Organic Matter

Organic farmers don't use artificial fertilisers, they use natural substances and processes instead.

1) Organic farmers put animal manure, compost and human sewage onto their land as fertilisers. The human sewage is heat-treated first to destroy harmful microbes.
2) Manure, sewage and compost all contain waste plant material, so they replace the elements that plants take out of the soil in the same way as a natural cycle would (see previous page).
3) Organic farmers also grow "green manure". Plants are grown on fields and then ploughed in and left to rot. Plants like clover are often used because they have nitrogen-fixing bacteria in their roots which add nitrates to the soil. The nitrates can then be used by the next crop.

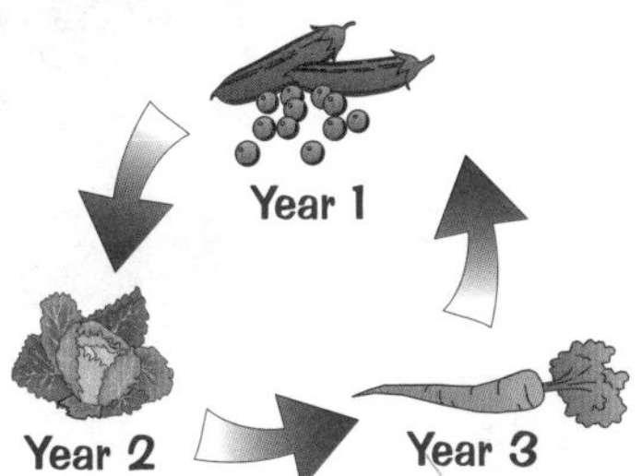

4) Organic farmers also use crop rotation to help keep their soil fertile. They grow different crops each year in a cycle — for example, peas might be grown one year, cabbage the next and carrots in year three. The peas contain nitrogen-fixing bacteria, so they help prepare the soil for the cabbage crop, which needs lots of nitrates. Carrots don't need much nitrogen, so they can still grow in the soil after the cabbage crop. And then it's back to the peas again.

2) Intensive Farming Relies on Artificial Fertilisers

1) Intensive farmers use man-made artificial fertilisers to put elements back into the soil.
2) Because artificial fertilisers are pure chemicals (i.e. they're not full of plant matter), it's easy to use just the right amount. Farmers can use small volumes of artificial fertilisers, because they contain much higher percentages of the elements the crops need than manure does.
3) Artificial fertilisers are spread on the ground as pellets or sprayed onto the crops as they grow.
4) The amount of each nutrient can be chosen to be exactly right for a particular plant's needs.

So that's intensive farming — now do some intensive revision...

There are a few important things to take away from this page. Without a doubt the most important thing is that when you're eating organic food, you could be eating elements from someone else's poo.

Pest Control

So you've got all the nutrients your crop needs in the soil and everything looks rosy (especially if you're growing roses). But just when you thought nothing could go wrong, bam... a plague of locusts. Oh no.

There are Different Ways to Deal With Pests and Disease

Pests and diseases are an absolute nightmare for farmers because they can seriously reduce crop yields.

1) Some pests, like aphids, are insects that eat crop plants.
2) Diseases like potato blight can damage or kill crop plants.

Luckily, most pests and diseases can be controlled. The methods used depend on the type of farming.

1 Organic Farming Uses Natural Biological Processes

Organic farmers aren't allowed to use man-made chemicals to deal with pests and diseases.

1) Pests can be controlled using natural predators — this is known as biological control. For example, ladybirds can be introduced into greenhouses to prey on greenfly pests.
2) Crop rotation (see previous page) is used to prevent the pests and disease causing organisms of one particular crop plant building up in an area.
3) Field edges are left grassy to encourage larger insects and other animals that feed on pests.
4) Varieties of plants that are best able to resist pests and diseases are chosen.
5) Natural pesticides can also be used. Some pesticides are completely natural, and as long as they're used responsibly they don't mess up the ecosystem.

2 Intensive Farming Relies on Chemicals

Intensive farmers spray their crops with man-made chemicals to destroy pests and diseases.

1) Chemical pesticides are more effective than organic methods. They usually kill all of the pests and disease-causing organisms, which organic methods can't.
2) This means a bigger yield of crops with fewer blemishes.
3) But it's not all good — the spraying leaves a chemical residue on the crop. This could harm humans eating the plants, as well as the pests.
4) Chemical pesticides kill indiscriminately — this means that not only pests are killed but also other organisms that could be beneficial.

Organic Farmers Have to Follow Certain Rules

It's illegal to sell food as organic if it hasn't really been grown that way.

1) The UK government has set national standards that have to be met by organic farmers, e.g. concerning the use of chemicals.
2) The national rules ban the use of virtually all artificial chemicals and set standards for the way that pests and diseases are controlled. The levels of pesticides and other artificial chemicals in the soil have to be below a certain level before a farm can be classed as organic.
3) The standards for meat that is classed as organic are just as strict. The animals must be allowed to move around freely, can only be fed on organic feed and can't be given artificial hormones to make them grow more quickly. They're not given medication unless it's really necessary, unlike intensively reared animals, which are often given antibiotics as a matter of course because infections are so common.

I'm arresting you on suspicion of carrot fraud...

Both types of farming have their own costs and benefits — for the general public, the environment, the farmers, etc. Organic farming is more expensive and you can't grow as much in one area, but it is more sustainable — it's less harmful to the environment and you don't have to manufacture chemicals.

Natural Polymers

When you think about polymers (if you ever do), you probably think of the man-made kind, like plastics. But as usual Mother Nature got there first. And her polymers don't choke seagulls and stuff either.

Carbohydrates and Proteins are Natural Polymers

Many of the chemicals found naturally in living things are long-chain molecules called polymers.

1) Polymers are large molecules formed by combining lots of smaller molecules (called monomers) in a regular pattern.
2) Complex carbohydrates are polymers built by linking together simple sugars like glucose.
3) Proteins are huge polymer molecules built by linking small molecules called amino acids.

Carbohydrates Consist of Carbon, Hydrogen and Oxygen

Carbohydrates are a group of compounds that include sugars (the monomers) and the polymers cellulose and starch. They're all made up of the same elements — carbon, hydrogen and oxygen.

1) The name 'carbohydrate' comes from the fact that they're basically made of carbon and water.
2) The simplest carbohydrate is the sugar glucose, $C_6H_{12}O_6$. This is the sugar that plants make from carbon dioxide and water, by photosynthesis.
3) Other carbohydrates like starch (for energy storage in plants) and cellulose are made by linking glucose molecules together in chains.

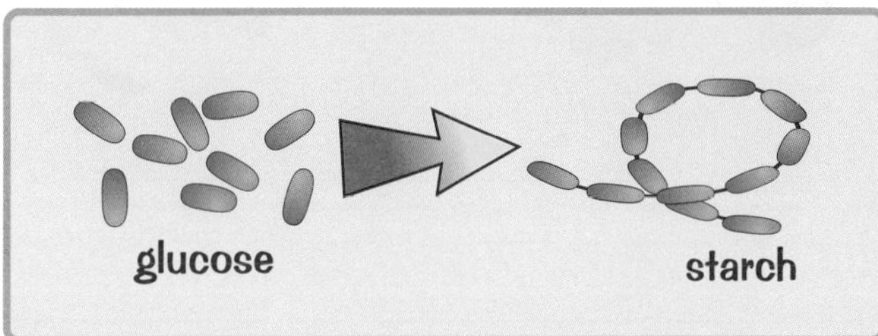

Proteins Contain Carbon, Hydrogen, Oxygen and Nitrogen

Proteins are a huge group of compounds. They're all made from amino acids and consist mostly of carbon, hydrogen, oxygen and nitrogen.

1) Plants produce their own amino acids, which are then linked together into long chains to form proteins.

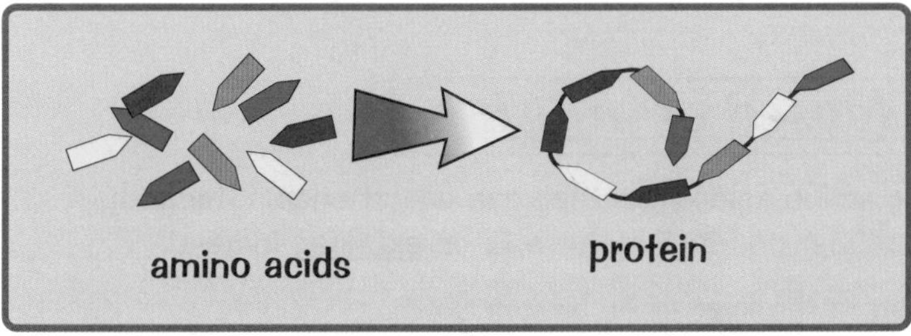

2) Animals have to get their amino acids from plants. They break down plant proteins into amino acids during digestion, and then join them back up in a new order in their cells to make new proteins.
3) Amino acids all have the same basic structure with slight variations.
4) There are only about 20 different amino acids used in the human body. It's the order they're connected in that makes each protein different.

Zzzzzzzzzzzzzzzz... What? Who? No, of course I wasn't asleep...

It's pretty mind-numbing this, isn't it? Still, dull pages are bound to crop up from time to time and you've just got to grit your teeth and learn them. If you're still awake please learn the elements that carbohydrates and proteins are made from — that's the key thing you need to know.

Digestion

You get food from animals and plants. But you don't want their carbohydrates and proteins, you want your own human-type ones. So first you break down the polymers you eat, and then you build new ones.

Digestion Involves Breaking Down Big Molecules

After you've chewed your food up and your stomach's had its turn at churning it up even further, it's still made up of quite big molecules, namely: starch, proteins and fats.

These are insoluble and too big to diffuse into the blood, so they're broken down in the small intestine into smaller, soluble molecules like glucose and amino acids. These can then move into the blood and be transported to all the cells of the body where they're needed.

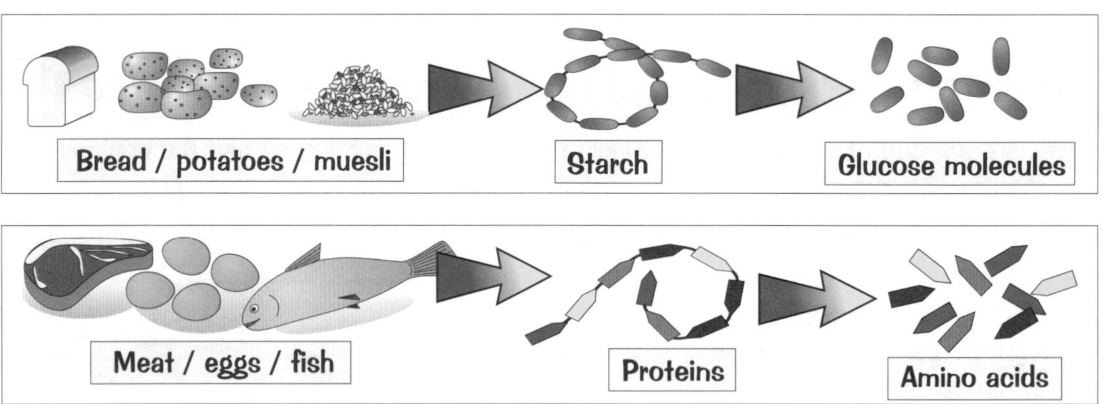

Amino Acids Are Built Up into New Proteins

1) As cells grow they need new proteins, which they make from the amino acids flowing past in the blood.
2) Each protein has a different sequence of amino acids. One way to think of it is like the letters of the alphabet — every word (protein) is a different arrangement of the same letters (amino acids).
3) Most of your body is made from proteins. They're the main part of your skin, hair, muscles and tendons.
4) Haemoglobin (the red stuff in red blood cells) is a protein called globin attached to a substance called haem. Haem contains iron and is the substance that carries oxygen around your body.

Excess Amino Acids Are Disposed of in Urine

Animals often eat protein that contains more amino acids than they can use at once. The body can't usually store amino acids and use them later on, so they have to be excreted instead.

1) Excess amino acids in the blood are taken to the liver to be broken down.
2) They are converted into a soluble substance called urea which is then released into the blood.
3) As the blood passes through your kidneys, the dissolved urea is removed and passes out of the body in the urine.

Digestion? I'm sure they told me this was chemistry...

Ah you see, it's not like the old days when biology was pondweed, chemistry was Bunsen burners and physics was light bulbs. Now that you're older, science is all overlapping and full of issues. It's more relevant to real life, of course — but sometimes I still kind of miss the Bunsen burners.

Insulin and Diabetes

Everyone knows that it's not healthy to eat loads of processed foods and to be overweight, and yet loads of us still eat junk. That's why more people are developing type 2 diabetes, and at a younger age too.

Diabetes is When The Body Can't Control Blood Sugar Levels

The body produces a hormone called insulin in the pancreas. Insulin helps glucose to move from the blood into body cells. This reduces blood glucose and gets the glucose where it needs to be to release its energy. Insulin also controls the storage of glucose. The amount of insulin you produce depends on your blood sugar level. The more sugar you've eaten and released into the blood, the more insulin is produced.

1) Processed foods often contain lots of sugar. For example, a can of cola contains 35 g of sugar (about 8 teaspoons). Sugar enters the bloodstream very quickly, making your blood sugar level shoot up. That's why it's better to eat complex carbohydrates (in brown bread, rice and cereals), which are broken down into sugar gradually.
2) Diabetes is an illness that occurs when the body can't control its blood sugar levels properly. There are two types of diabetes — type 1 and type 2.
3) Poor diet and obesity increase your risk of developing type 2 (but not type 1) diabetes.

The Two Types of Diabetes Cause Problems in Different Ways

The two types of diabetes develop for different reasons and are treated in different ways.

TYPE 1

1) Type 1 diabetes usually develops in young people and happens when the pancreas stops producing insulin, for reasons that aren't fully understood yet.
2) Because the body can't produce insulin, it can't remove sugar from the blood and store it. Blood sugar levels can become so high that they damage the body, possibly even causing coma and death.
3) Type 1 diabetes is treated by daily injections of insulin and by controlling the diet to help control the blood sugar levels. It's important that diabetics don't miss meals, or the insulin they've injected could make their blood sugar levels drop too low, which can also be dangerous.

TYPE 2

1) Type 2 diabetes usually affects older people and symptoms develop gradually. They include weight loss, needing to wee often and tiredness. It can develop if the body stops making enough insulin because it can't respond properly to high blood sugar levels. Or the body might be producing enough insulin, but stops responding to it normally. Both these things are linked to a poor diet and obesity.
2) Type 2 diabetes is controlled by improving the diet, losing weight and exercising regularly. Exercise can help to keep the blood sugar level stable and helps prevent obesity. Sometimes insulin is also taken.
3) Type 2 diabetes is becoming more and more common in young people. This is because of increasing obesity due to poor diet and lack of exercise.

Just 8 teaspoonfuls of sugar leads to obesity and diabetes...

One thing that they're keen on in science these days is risk, and this is just the kind of place they'd spring it on you. For example, what risks are associated with being obese? Why might people accept the risk involved in having a poor diet? (Think about stuff like convenience, cost, wide availability of processed foods.) You must be able to discuss personal choices in terms of balancing risks and benefits.

Harmful Chemicals in Food

You need to eat to live. But some foods can do you more harm than good...

Some Foods Are Dangerous Naturally...

Some plants contain poisonous chemicals, some are poisonous unless cooked properly and some cause allergic reactions in some people.

1) A lot of mushrooms are poisonous. Only a few are deadly, but many can cause stomach upsets.
2) Cassava, a plant found in South America, has a floury root that's widely eaten. In the root are compounds that produce lethal cyanide in the liver if eaten, but luckily it also contains a substance that destroys these compounds if it's mixed with them. This is normally done by either chopping up the roots and boiling them, or by fermenting them.
3) As many as one in 200 people are allergic to proteins found in peanuts and other nuts. In extreme cases this can be fatal, but more often the symptoms are a rash and swelling of the mouth and throat.
4) Gluten, a protein found in wheat, rye and barley, can cause allergic reactions in some people. Symptoms vary and may include rashes and swellings, stomach pain and vomiting, diarrhoea and bloating or even breathing problems. People with this allergy have to eat a low gluten or gluten-free diet.

...Others Contain Chemicals Left Over From Farming...

Pesticides and herbicides used in farming can be present in small amounts on fruit and vegetables.

1) These residues are usually in very small amounts and can be removed by careful washing and peeling.
2) The long-term effect on health of eating small amounts of these chemicals is not fully understood, so many people now do their best to avoid them. A recent study has linked residues with an increased risk of Parkinson's disease — a degenerative brain disease.

...Others Develop Dangerous Chemicals During Storage...

Some foods can contain harmful chemicals if they're not stored properly.

1) Nuts, cereals and dried fruit can become contaminated by a poisonous substance called aflatoxin.
2) Aflatoxin is produced by a mould that can grow on these foods if they're not stored correctly.
3) Aflatoxin can cause liver cancer.
4) Bird seed is a particular problem, as it doesn't have to meet the same standards as nuts and seeds for human consumption. Birds and other animals can be killed if they eat contaminated seed.

...And Cooking Can Form Other Dangerous Chemicals

Burning foods can produce dangerous chemicals.

1) These chemicals (called HAs and PAHs) are usually formed when food is cooked at high temperatures (over 150 °C or so).
2) They can be formed in a whole range of different foods, including meat and fish.
3) HAs and PAHs have been shown to cause cancer in animals by altering their DNA.
4) The amounts of these chemicals produced can be reduced by choosing lean meat or fish, not letting flames touch the food and by cooking it at a lower temperature for longer.

Think twice before you stick another shrimp on the barbie...

Here's that risk and benefit stuff again — scientists are always coming up with new ways that various foods are bad for us. But if you spend too much time analysing all those risks, you'd never eat at all.

Food Additives

You're not finished with food issues yet, I'm afraid. Here are all the chemicals that are put in on purpose.

Lots of Foods Contain Additives These Days

1) Food colours (should) make food look more appetising. They're often used in sweets and soft drinks. A place you might not expect to find food colouring is in mushy peas, which contain a green dye.

2) Flavourings, fairly obviously, are added to foods to give them a new taste — e.g. adding an orange flavour to a soft drink. They could be extracted from a natural substance or made artificially. Flavour enhancers are a bit different — they bring out the taste and smell of food without adding a taste of their own (they're not flavourings as such). They're often added to ready meals.

3) Artificial sweeteners (like saccharin and aspartame) are used in things like diet foods and drinks instead of sugar. They taste sweeter than sugar so you don't have to use as much, making the products lower in calories.

4) Antioxidants stop some foods from going off when they react with oxygen. Oxygen can turn the fat in food into nasty-smelling and nasty-tasting substances — e.g. butter goes rancid when it's exposed to the air. Antioxidants are added to foods that contain fat or oil, e.g. sausages, to stop them reacting with oxygen.

5) Preservatives are added to many foods to prevent the growth of harmful microbes. The food can then be stored for longer before it goes off. A very common preservative is sodium benzoate.

Emulsifiers and Stabilisers Help Oils and Water Mix

1) Oil and water naturally separate into two layers with the oil floating on top of the water — they don't "want" to mix. Emulsifiers help to stop the two liquids in an emulsion separating out — you'll find them in things like mayonnaise.
2) Stabilisers are added to foods to help emulsions stay mixed and to thicken them. They're added to lots of foods, e.g. tomato sauce, ice cream and many other desserts.

The Use of Food Additives is Regulated

Food additives have to pass a safety test before they can be used.

1) In the EU (including the UK), all food additives have to pass a safety test and are then given an E number. Even oxygen has an E number, because it's used with gas-packed vegetables.
2) The standards set by the EU are different from those used in other countries. Some substances are allowed in the EU but not in other countries and vice versa.
3) Despite these safety tests, some additives are still thought to cause health problems in some people. For example:

- Some artificial food colourings have been linked with allergies and hyperactivity.
- Sulfur dioxide in dried fruit has been linked with asthma.
- Artificial sweeteners, e.g. aspartame, have been linked with hyperactivity and behavioural problems.

OK, enough — this is putting me off my chemical-filled lunch...

E numbers are something else that have had a very bad press lately. And fair enough, luminous coloured sweeties that have kids bouncing off the walls are probably a bit unnecessary. But some of the chemicals that are added are really useful, like preservatives that stop harmful microbes growing.

Keeping Food Safe

Reducing the risk of a hazard, e.g. a pesticide residue on food, costs money.
Governments have to balance the risks and costs to reach an acceptable level of risk.

Foods Can Be Made Safer but Not Risk-free

Everything you do carries a certain amount of risk, even something as simple as eating a meal. New technology based on scientific advances introduces new risks, and these have to be limited.

1) For example, scientists genetically modify some crops to give them new characteristics, like better yields and resistance to diseases. However, some people are worried that this new technology might not be safe, and so research into GM crops is strictly regulated. Food containing material from GM organisms must be clearly labelled so that consumers can avoid it if they want.
2) Scientists are also responsible for many of the chemicals sprayed on or added to foods (herbicides, pesticides, flavourings, etc.). These have their advantages, but they also pose risks (see p.24). Governments and other organisations evaluate these risks and impose controls accordingly (see below).
3) No food can ever be guaranteed to be completely safe. There are too many stages in the food chain and too many different ways that a food could be contaminated. New allergies and food intolerances can develop at any point in a person's life. You simply have to accept a small amount of risk, and take what steps you can to make sure it's as low as possible.
4) For example, a lot can be done to reduce the potential risks of eating meat:
   - The farmer should care for the animals properly.
   - The animals must be slaughtered in a hygienic environment.
   - The butcher must keep the meat in a clean and cold place.
   - Members of the public should store and prepare the meat in the safest way.

   The government aims to ensure all this happens by licensing places like slaughterhouses, setting out guidelines, enforcing laws and educating the public.

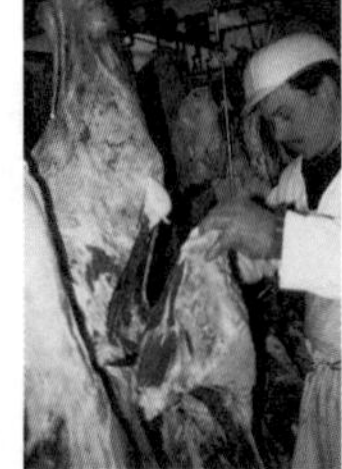

The 'precautionary principle' is an idea used by governments and individuals to help limit risks. For example, if you're not sure about some food, say unlabelled meat on a market stall in the sun, you shouldn't buy it unless you're prepared to accept the risk that it might be unsafe, however cheap it is. Basically, if you're not sure about something, and you know it could potentially cause serious harm, you'd best avoid it.

Scientists Decide Safe Levels of Chemicals in Food

Foods are regulated to make sure they don't pose a significant risk to health.

1) Potentially dangerous chemicals get into food from a variety of sources and at every stage in the food supply chain — from pesticides in the fields, to processing, packaging and storage.
2) Scientific advisory committees, which aren't connected to the food industry, carry out risk assessments to help set safe limits for the levels of chemicals allowed in food.
3) The Food Standards Agency (FSA) is an independent food safety watchdog. It offers advice to consumers and food producers on all aspects of food safety, labelling, diet, farming and hygiene.
4) The FSA also checks that legislation on these issues is being followed properly, for example by supporting food sampling programmes. These are carried out regularly by local authorities and involve testing various foods from different sources to make sure they're safe to eat.

I laugh in the face of danger — see, I haven't washed this pear...

I told you — these examiners love the whole idea of risk. In their spare time I expect they all go skydiving. This topic is an ideal way for them to introduce ideas about risk, but it's not the only place it could come up in an exam, so make sure you understand the general ideas behind these examples.

Eating Healthily

So the Government and various organisations are keeping a beady eye on the companies and individuals producing our food, making sure they don't poison us. But the Government can't be responsible for everything you put in your mouth.

Individual Choices Can Help Make Food Safer

Everyone should be aware of the possible harmful effects of the various chemicals in their food. Individuals who want to reduce their exposure to potentially harmful chemicals can take several steps:

1) Choose food produced in a way that minimises the amount of artificial chemicals applied to it. This usually means eating organic food.
2) Wash the food carefully or peel it before eating it.
3) Store and cook the food in the way recommended on the packaging.

People can also use the labels on food to find out more about the consequences for their health.

- Food labels give detailed information about things like the amounts of each type of fat, and sometimes whether this is high, medium or low compared with other foods.

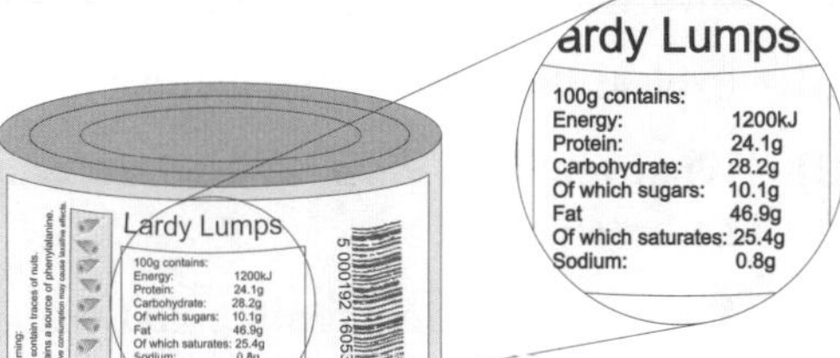

The label should tell you:

- How and where the food was produced.
- What it contains and whether it contains substances that people might be allergic to.

Many labels also give you an idea of the recommended daily amounts of different substances you should be eating, and how much of that daily amount the product contains.

Choosing Food Isn't Just About What it Tastes Like

1) People don't all eat exactly the same diet. There's so much choice that two people's diets might be completely different — one person might avoid fruit and vegetables altogether, and the other might be a vegan who eats a diet of organic foods.
2) More and more people are becoming concerned about the food they eat and many are turning to organic food. Organic food is often seen as safer because fewer artificial chemicals are used in producing it, and some people see it as more natural and nutritious. For these people, the potential risks posed by eating food from intensive farms don't seem worth the benefits.
3) But there are benefits to intensively grown foods. They're cheaper (an important factor for many people). They often look more attractive, and there tends to be more choice.
4) Processed ready meals are quick and convenient. For some these benefits outweigh the potential risks posed by high sugar, fat and salt contents, and other additives in the food.
5) All food you buy is subject to lots of government regulations, so unless you have an allergy to one of the ingredients, it's unlikely to poison you. This is enough for some people to dismiss all the risks of an unhealthy diet and eat whatever they like.

Your diet is a personal choice, and each person balances the risks and benefits of eating different foods for themselves based on their own opinions. Advice from the Government or their doctor, and what they've heard from other people and from the media might all influence their final decision.

I choose the seefood diet — when I see food I eat it...

Isn't it crazy that even though there's more choice these days, and more and more people are eating free range, organic, local, and whatever, obesity is still one of the biggest causes of death in countries like the USA and UK. That's personal choice for you. Put down that pie and get out on your bike.

Revision Summary for Module C3

Okay, I'm sure you know what to do by now — and if by some chance you've forgotten, that great big list of questions should give you a clue. Go through and try them all, making a note of any you can't do. Then go back through the section and find the answers to the ones you were stuck on. And I warn you, if you don't try these questions, whenever you try to grow rhubarb it will not sprout, and whenever you make custard it will turn out lumpy.

1) How is dead animal and plant matter turned into compounds that plants can use?
2) Give three ways that nitrates are added to the soil in the nitrogen cycle.
3) Give two ways that nitrates are removed from the soil in the nitrogen cycle.
4) Explain why the nitrogen cycle does not happen in a field of crops.
5) Give three ways that organic farmers can replace the nutrients that are lost from their soil.
6) Give two advantages and two disadvantages of using artificial fertilisers on crops.
7) Give two ways that an organic farmer can limit the number of diseases affecting his or her crops.
8) Explain the advantages and disadvantages of using chemical pesticides to kill pests.
9) What must a farmer who grows crops and produces animals for meat do for the food to be classed as organic?
10) Name two carbohydrate polymers.
11) Which element is found in proteins but not in carbohydrates?
12) Name the monomer molecules that make up a protein polymer.
13) Why can the starch in food not be absorbed straight into the blood?
14) Where are excess amino acids broken down in the body? How are the products of this process excreted?
15) How is type 1 diabetes usually treated?
16) Explain why type 2 diabetes is increasing in young people.
17) Explain why the South American plant, cassava, should not be eaten raw.
18) What is aflatoxin? What problem can it cause if eaten by humans?
19) What health problem might be caused by eating burnt foods?
20) Why are the following added to food? a) sodium benzoate b) saccharin
21) Ice cream contains two different kinds of chemical to prevent it separating out — what are they?
22) Give two examples of problems that have been linked to additives passed as safe to use by the EU.
23)*Why can no food ever be guaranteed to be completely safe?
24) Explain what the 'precautionary principle' is.
25) Explain how scientific advisory committees and the FSA limit the risks from food.
26) Suggest three ways that individuals can limit the risks from their food.
27) What kinds of information can consumers get from food labels?
28)*Outline some of the risks and benefits of eating intensively produced vegetables.

* Answers on page 92.

Atoms

Atoms are the building blocks of everything — and I mean everything. They're amazingly tiny — you can only see them with an incredibly powerful electron microscope.

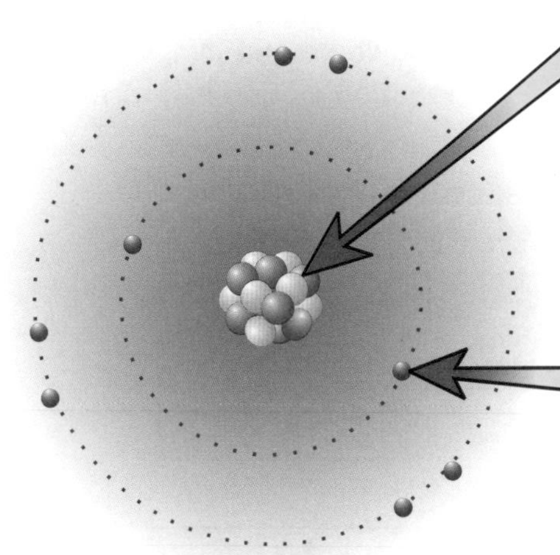

The Nucleus

1) It's in the middle of the atom.
2) It contains protons and neutrons.
3) It has a positive charge because of the protons.
4) Almost the whole mass of the atom is concentrated in the nucleus.
5) But size-wise it's tiny compared to the rest of the atom.

The Electrons

1) Move around the nucleus.
2) They're negatively charged.
3) They're tiny, but they cover a lot of space.
4) The volume of their orbits determines how big the atom is.
5) They have virtually no mass.
6) They're arranged in shells around the nucleus.
7) These shells explain the whole of Chemistry.

Know Your Particles...

1) Protons are heavy and positively charged.
2) Neutrons are heavy and neutral.
3) Electrons are tiny and negatively charged.

PARTICLE	MASS	CHARGE
Proton	1	+1
Neutron	1	0
Electron	0.0005	-1

(Electron mass is often taken as zero.)

Number of Protons Equals Number of Electrons

1) Neutral atoms have no charge overall.
2) The charge on the electrons is the same size as the charge on the protons — but opposite.
3) This means the number of protons always equals the number of electrons in a neutral atom.
4) If some electrons are added or removed, the atom becomes charged and is then an ion.
5) The number of neutrons isn't fixed but is usually about the same as the number of protons.

Each Element has a Different Number of Protons

1) It's the number of protons in an atom that decides what element it is.
For example, any atom of the element helium will have 2 protons — and any atom with 2 protons will be a helium atom.

> Atoms of the same element all have the same number of protons — and atoms of different elements will have different numbers of protons.

2) Elements all have different properties from each other due to differences in their atomic structure.

Number of protons = number of electrons...

This stuff might seem a bit useless at first, but it should be permanently engraved into your mind. If you don't know these basic facts, you've got no chance of understanding the rest of Chemistry. So learn it now, and watch as the Universe unfolds and reveals its timeless mysteries to you...

Balancing Equations

Every time you write an equation you need to make sure it balances rather than skate over it.

Atoms aren't Lost or Made in Chemical Reactions

1) Remember that during chemical reactions things don't appear out of nowhere and things don't just disappear. You still have the same atoms at the end of a chemical reaction as you had at the start. They're just arranged in different ways.
2) Balanced symbol equations show the atoms at the start (the reactant atoms) and the atoms at the end (the product atoms) and how they're arranged. For example:

Word equation:	sodium	+	chlorine	$\rightarrow$	sodium chloride
Balanced symbol equation:	2Na	+	Cl_2	$\rightarrow$	2NaCl
	Na Na		Cl Cl		Na Cl Na Cl

Balancing the Equation — Match Them Up One by One

1) There must always be the same number of atoms on both sides — they can't just disappear.
2) You balance the equation by putting numbers in front of the formulas where needed. For example...

$$Na + H_2O \rightarrow NaOH + H_2$$

The formulas are all correct but the numbers of some atoms don't match up on both sides. You can't change formulas like H_2O to H_3O. You can only put numbers in front of them:

Method: Balance Just ONE Type of Atom at a Time

The more you practise, the quicker you'll get, but all you do is this:

1) Find an element that doesn't balance and pencil in a number to try and sort it out.
2) See where it gets you. It may create another imbalance — if so, just pencil in another number and see where that gets you.
3) Carry on chasing unbalanced elements and it'll sort itself out pretty quickly.

I'll show you. In the equation above you'll soon notice we're short of H atoms on the LHS (Left-Hand Side).

1) The only thing you can do about that is make it $2H_2O$ instead of just H_2O:
$$Na + 2H_2O \rightarrow NaOH + H_2$$
2) But that now causes too many H atoms and O atoms on the LHS, so to balance that up you could try putting 2NaOH on the RHS (Right-Hand Side). You'll then need to put 2 in front of the Na on the LHS to balance the number of Na atoms:
$$2Na + 2H_2O \rightarrow 2NaOH + H_2$$
3) And suddenly there it is! Everything balances.

State Symbols Tell You What Physical State It's In

These are easy enough, so make sure you know them — especially aq (aqueous).

(s) — Solid (l) — Liquid (g) — Gas (aq) — Dissolved in water

E.g. $2K(s) + Cl_2(g) \rightarrow 2KCl(s)$

Balancing equations — weigh it up in your mind...

Remember what those numbers mean: A number in front of a formula applies to the entire formula. So, $3Na_2SO_4$ means three lots of Na_2SO_4. The little numbers in the middle or at the end of a formula only apply to the atom or brackets immediately before. So the 4 in Na_2SO_4 just means 4 O's, not 4 S's.

Line Spectrums

Colour isn't just to do with art — you've got to learn about it in Chemistry too.

Some Elements Emit Distinctive Colours When Heated

1) When heated, some elements produce flames with a distinctive colour.
   For example:

   (i) Lithium, Li, produces a red flame.
   (ii) Sodium, Na, produces a yellow/orange flame.
   (iii) Potassium, K, produces a lilac flame.

2) All the different colours seen in fireworks are due to the colours produced by different elements — it's a great bit of chemistry.
3) What's more... these colours also help chemists to identify a metal in a compound. Just put a little bit of the substance into a blue Bunsen flame, and see what colour's produced.

Each Element Gives a Characteristic Line Spectrum

1) When heated, the electrons in an atom are excited, and release energy as light.
2) The wavelengths emitted can be recorded as a line spectrum.
3) Different elements emit different wavelengths of light. This is due to each element having a different electron arrangement (see p.37).
4) So each element has a different pattern of wavelengths, and a different line spectrum.
5) This means that line spectrums can be used to identify elements.
6) The practical technique used to produce line spectrums is called spectroscopy.

A line spectrum for an element will look something like this:

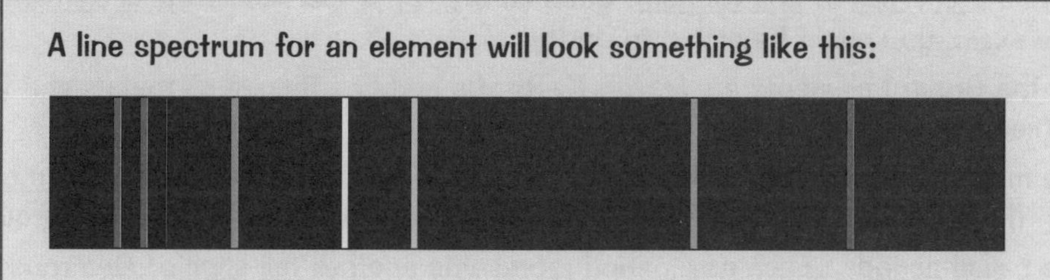

Line Spectrums Have Identified New Elements

New practical techniques (e.g. spectroscopy) have allowed scientists to discover new elements. Some of these elements simply wouldn't have been discovered without the development of these techniques.

- Caesium and rubidium were both discovered by their line spectrum.
- Helium was discovered in the line spectrum of the Sun.

Spectroscopy — it's a flaming useful technique...

These are quite nifty ways of identifying elements. Maybe more importantly, we've been able to discover elements that otherwise might still be unknown to us using these techniques. I know it's a bit tricky, but make sure you understand the details on this page.

The Periodic Table

The periodic table is a chemist's bestest friend — start getting to know it now... seriously...

The Periodic Table Puts Elements with Similar Properties Together

1) There are 100ish elements, which all materials are made of. If it wasn't for the periodic table organising everything, you'd have a heck of a job remembering all those properties. It's ace.
2) It's laid out in order of increasing proton number.
3) It's a handy tool for working out which elements are metals and which are non-metals. Metals are found to the left and non-metals to the right.

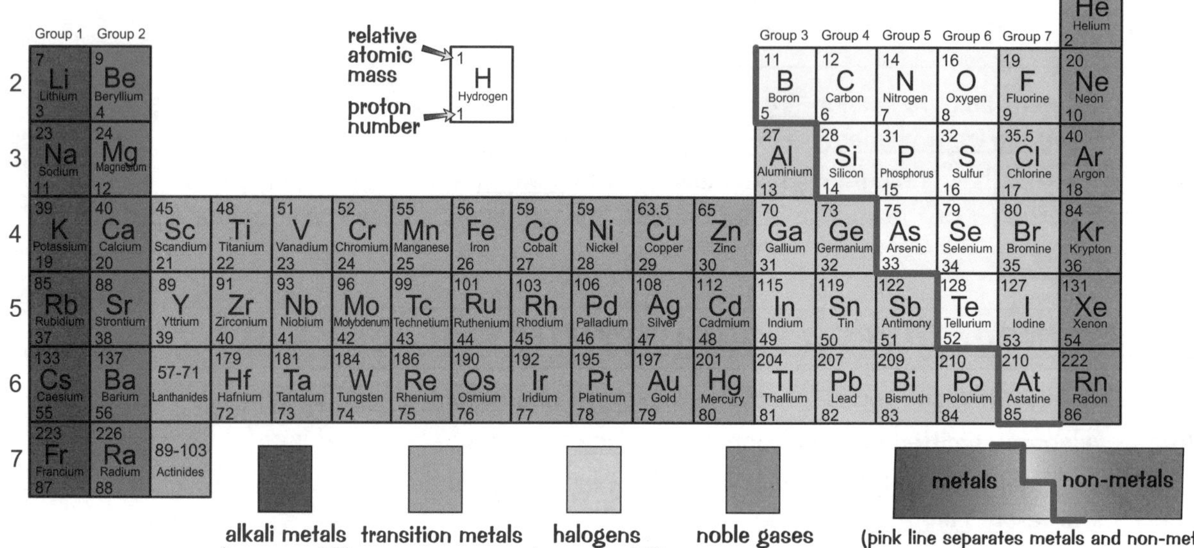

4) Elements with similar properties form columns.
5) These vertical columns are called groups and Roman numerals are often (but not always) used for them.
6) If you know the properties of one element, you can predict properties of other elements in that group — and in the exam, you might be asked to do this.
7) For example the Group 1 elements are Li, Na, K, Rb, Cs and Fr. They're all metals and they react in a similar way (see page 38).
8) You can also make predictions about reactivity. E.g. in Group 1, the elements react more vigorously as you go down the group. And in Group 7, reactivity decreases as you go down the group.
9) The rows are called periods. Each new period represents another full shell of electrons (see page 37).

You Can Get Loads of Information from the Periodic Table

By looking at the table, you can immediately find out:

1) The name and symbol of each element.
2) The proton number of each element — this tells you how many protons there are in the nucleus.
3) The relative atomic mass of each element — this tells you the total number of protons and neutrons there are in the nucleus.

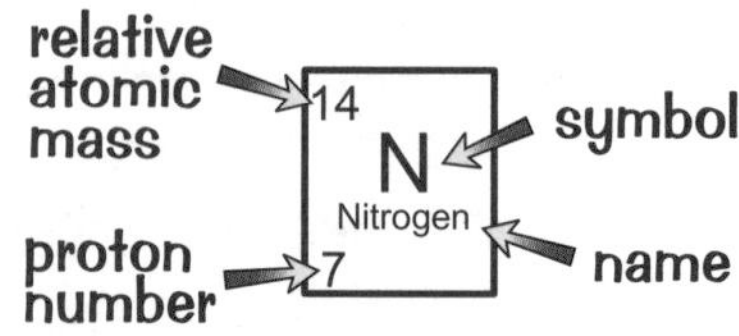

I'm in a chemistry band — I play the symbols...

Scientists keep making new elements and feeling well chuffed with themselves. The trouble is, these new elements only last for a fraction of a second before falling apart. You don't need to know the properties of each group of the periodic table, but if you're told, for example, that fluorine (Group 7) forms two-atom molecules, it's a fair guess that chlorine, bromine, iodine and astatine do too.

Electron Shells

Electron shells... the orbits electrons zoom about in.

Electron Shell Rules:

1) Electrons always occupy shells (sometimes called energy levels).
2) The lowest energy levels are always filled first.
3) Only a certain number of electrons are allowed in each shell:
   1st shell: 2   2nd Shell: 8   3rd Shell: 8
4) Atoms are much happier when they have full electron shells.
5) In most atoms the outer shell is not full and this makes the atom want to react.
6) An element's electron arrangement determines its chemical properties. (See pages 38 and 39.)

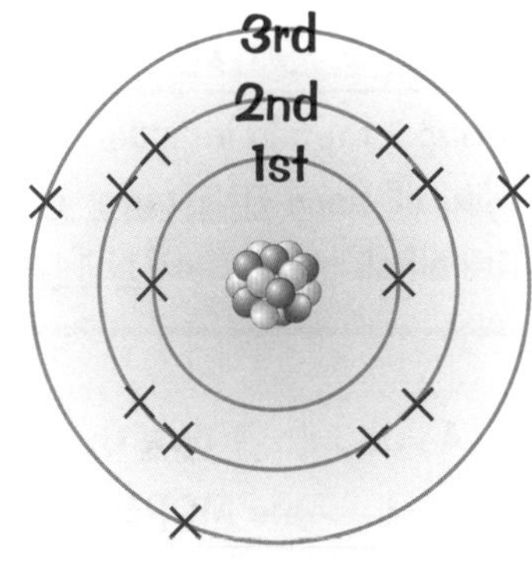

3rd shell still filling

Working Out Electron Configurations

You need to know the electron configurations for the first 20 elements in the periodic table. They're shown in the diagram below — but they're not hard to work out.
For a quick example, take nitrogen. Follow the steps...

1) The periodic table tells you that nitrogen has seven protons... so it must have seven electrons.
2) Follow the 'Electron Shell Rules' above. The first shell can only take 2 electrons and the second shell can take a maximum of 8 electrons.
3) So the electron configuration for nitrogen must be 2, 5 — easy peasy.
4) Now you try it for argon.

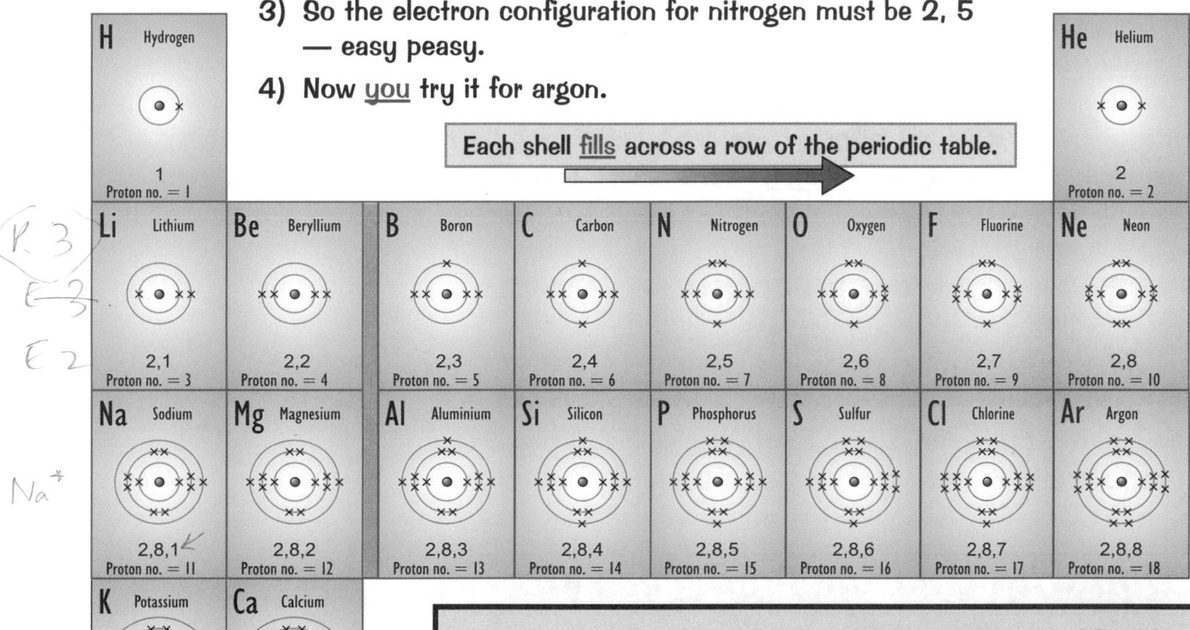

Answer: To calculate the electron configuration of argon, follow the rules. It's got 18 protons, so it must have 18 electrons. The first shell must have 2 electrons, the second shell must have 8, and so the third shell must have 8 as well. It's as easy as 2, 8, 8.

One little duck and two fat ladies — 2, 8, 8...

You need to know enough about electron shells to draw out that whole diagram at the bottom of the page without looking at it. Obviously, you don't have to learn each element separately — just learn the pattern. Cover the page: using a periodic table, find the atom with the electron configuration 2, 8, 6.

Group 1 — Alkali Metals

Alkali metals are all members of the same group — Group 1.
So, yes, you've got it — that means they'll all have similar properties.

Group 1 Metals are Known as the 'Alkali Metals'

1) Group 1 metals include lithium, sodium and potassium... know these names really well.
2) They all have ONE outer electron. This makes them very reactive and gives them all similar properties.
3) The alkali metals are shiny when freshly cut, but quickly tarnish in moist air.

As you go DOWN Group 1, the alkali metals:

1) become MORE REACTIVE (see below)
   ...because the outer electron is more easily lost, because it's further from the nucleus.
2) have a HIGHER DENSITY
   ...because the atoms have more mass.
3) have a LOWER MELTING POINT
4) have a LOWER BOILING POINT

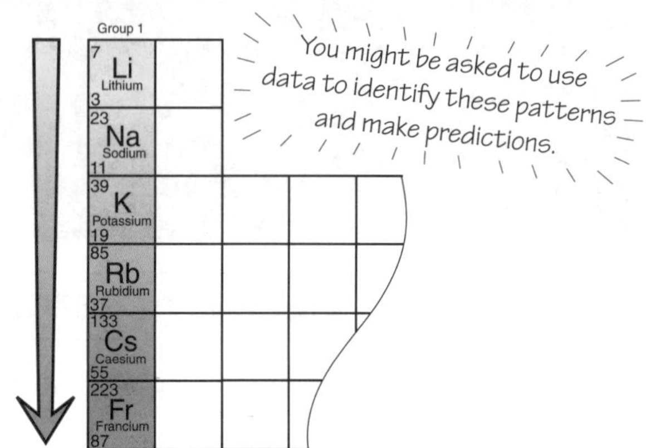

Reaction with Cold Water Produces Hydrogen Gas

1) When lithium, sodium or potassium are put in water, they react very vigorously.
2) They move around the surface, fizzing furiously.
3) They produce hydrogen. Potassium gets hot enough to ignite it. If it hasn't already been ignited by the reaction, a lighted splint will indicate hydrogen by producing the notorious "squeaky pop" as it ignites.
4) The reaction makes an alkaline solution — this is why Group 1 is known as the alkali metals.
5) A hydroxide of the metal forms, e.g. sodium hydroxide (NaOH), potassium hydroxide (KOH) or lithium hydroxide (LiOH).

$$2Na_{(s)} + 2H_2O_{(l)} \rightarrow 2NaOH_{(aq)} + H_{2(g)}$$
$$2K_{(s)} + 2H_2O_{(l)} \rightarrow 2KOH_{(aq)} + H_{2(g)}$$

6) This experiment shows the relative reactivities of the alkali metals. The more violent the reaction, the more reactive the alkali metal is.

Reaction with Chlorine Produces Salts

1) Alkali metals react vigorously with chlorine.
2) The reaction produces colourless crystalline salts, e.g. lithium chloride (LiCl), sodium chloride (NaCl) and potassium chloride (KCl).

$$2Na_{(s)} + Cl_{2(g)} \rightarrow 2NaCl_{(s)}$$
$$2K_{(s)} + Cl_{2(g)} \rightarrow 2KCl_{(s)}$$

Notorious Squeaky Pop — a.k.a. the Justin Timberlake test...

Alkali metals are ace. They're so reactive you have to store them in oil — because otherwise they'd react with the water vapour in the air. AND they fizz in water and explode and everything. Cool.

Group 7 — Halogens

The 'trend thing' happens in Group 7 as well — no surprise there.

Group 7 Elements are Known as the Halogens

1) Group 7 elements include chlorine, bromine and iodine... remember these names.
2) They all have SEVEN outer electrons. This makes them very reactive and gives them similar properties.
3) Chlorine's good because it kills bacteria. It's used in bleach and swimming pools. The other halogens act in a similar way.

As you go DOWN Group 7, the halogens:

1) become LESS REACTIVE (see below)
   ...because the outer electrons are further from the nucleus and so additional electrons are attracted less strongly.
2) have a HIGHER MELTING POINT
3) have a HIGHER BOILING POINT

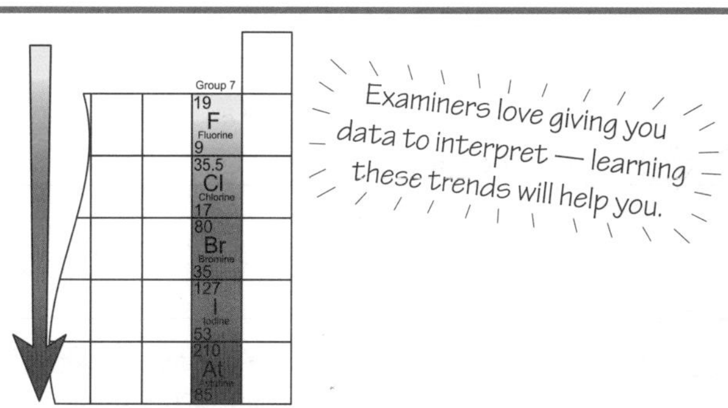

The Halogens are All Non-metals with Coloured Vapours

Fluorine is a very reactive, poisonous yellow gas at room temperature.
Chlorine is a fairly reactive, poisonous dense green gas at room temperature.
Bromine is a dense, poisonous, red-brown volatile liquid at room temperature and forms a red-brown gas.
Iodine is a dark grey crystalline solid at room temperature or a purple vapour.

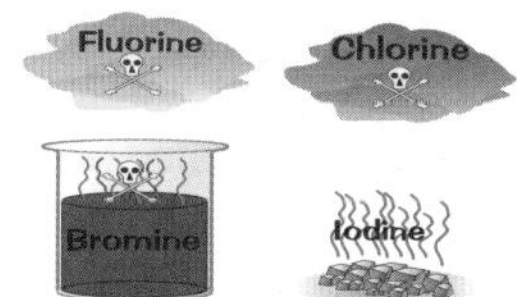

The halogens go from gases to solids down the group — this shows the trend in melting and boiling points.

They All Form Diatomic Molecules Which are Pairs of Atoms:

Cl_2 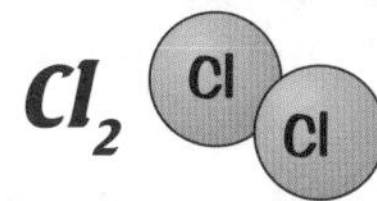

Br_2 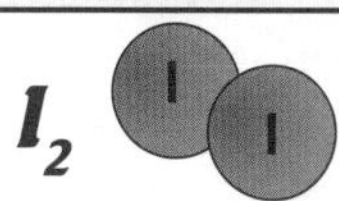

I_2

More Reactive Halogens Will Displace Less Reactive Ones

A displacement reaction is where a more reactive element 'pushes out' (displaces) a less reactive element from a compound.

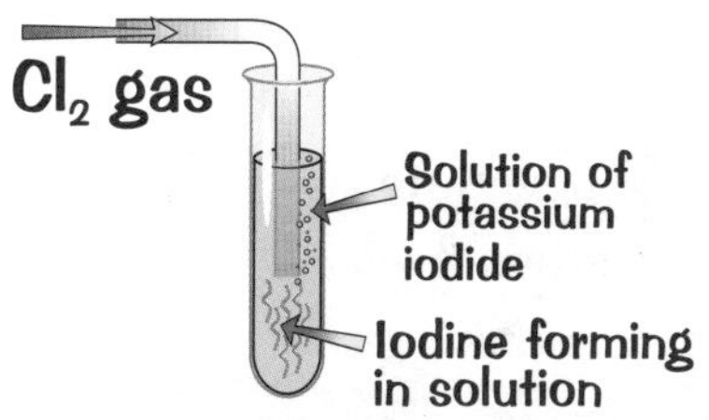

Chlorine is more reactive than iodine. So chlorine reacts with potassium iodide solution to form potassium chloride, and the iodine is left in solution.
Chlorine can also displace bromine from solutions of bromides.
Bromine will displace iodine because of the trend in reactivity.

$$Cl_2(g) + 2KI(aq) \rightarrow I_2(aq) + 2KCl(aq)$$
$$Cl_2(g) + 2KBr(aq) \rightarrow Br_2(aq) + 2KCl(aq)$$

These displacement reactions can be used to determine the relative reactivity of the halogens.

They're great, the halogens — you have to hand it to them...

The halogens are another group from the periodic table, and just like the alkali metals (p.38) you've got to learn their trends and the equations on this page. Learn them, cover up the page, scribble, check.

Laboratory Safety

You Need to Learn the Common Hazard Symbols...

Lots of chemicals can be bad for you or dangerous in some way.
These hazard symbols might just save your skin...

Oxidising

Provides oxygen which allows other materials to burn more fiercely.
Example: Liquid oxygen.

Harmful

Like toxic but not quite as dangerous.
Example: Copper sulfate.

Highly Flammable

Catches fire easily.
Example: Petrol.

Irritant

Not corrosive but can cause reddening or blistering of the skin.
Examples: Bleach, children, etc.

Toxic

Can cause death either by swallowing, breathing in, or absorption through the skin.
Example: Hydrogen cyanide.

Corrosive

Attacks and destroys living tissues, including eyes and skin.
Example: Concentrated sulfuric acid.

... And Know How to Work Safely with Dangerous Chemicals

Some of the elements you've come across so far in this section, like the alkali metals (see p.38) and the halogens (see p.39), are pretty dangerous. There are certain safety precautions that need to be followed when using these chemicals. The usual 'wear safety specs' goes without saying...

Alkali Metals

1) The Group 1 elements are really reactive and can combust spontaneously.
2) If they come into contact with water vapour in the air there can be a violent reaction, depending on how much alkali metal is present — so they're stored under oil to prevent this. Make sure there's a fire extinguisher handy...
3) Alkali metals should never be touched with bare hands — the sweat on your skin is enough to cause a reaction that will produce lots of heat and a corrosive hydroxide. Not pleasant.
4) Every piece of apparatus used in an experiment needs to be kept completely dry.
5) The alkaline solutions they form are corrosive and may cause blistering — it's important they don't touch the eyes or the skin.

Halogens

1) The Group 7 elements are also harmful. Chlorine and iodine are both very toxic.
2) Fluorine is the most reactive halogen — it's too dangerous to use inside the lab.
3) Liquid bromine is corrosive and so contact with the skin must be avoided.
4) Halogens have poisonous vapours that irritate the respiratory system and the eyes. They must be used inside a fume cupboard so that you don't breathe in the fumes.

No, it means 'oxidising' — not a guy with a wacky hairstyle...

The stuff on this page is all pretty important, not just for passing your exam but also for when you're doing experiments with chemicals in the lab. Make sure you know what the hazard symbols that appear on the containers of chemicals mean — they're not just there to look pretty...

Ionic Bonding

This stuff's a bit tricky, but keep at it and you'll be bonding with it in no time...

Ionic Bonding — Swapping Electrons

In ionic bonding, atoms lose or gain electrons to form charged particles (called ions) which are then strongly attracted to one another (because of the attraction of opposite charges, + and –).

A Shell with Just One Electron is Well Keen to Get Rid...

All the atoms over at the left-hand side of the periodic table, e.g. sodium, potassium, calcium etc., have just one or two electrons in their outer shell. And they're pretty keen to get shot of them, because then they'll only have full shells left, which is how they like it. So given half a chance they do get rid, and that leaves the atom as an ion instead. Now ions aren't the kind of things that sit around quietly watching the world go by. They tend to leap at the first passing ion with an opposite charge and stick to it like glue.

A Nearly Full Shell is Well Keen to Get That Extra Electron...

On the other side of the periodic table, the elements in Group 6 and Group 7, such as oxygen and chlorine, have outer shells that are nearly full. They're obviously pretty keen to gain that extra one or two electrons to fill the shell up. When they do of course they become ions (you know, not the kind of things to sit around) and before you know it, pop, they've latched onto the atom (ion) that gave up the electron a moment earlier. The reaction of sodium and chlorine is a classic case:

The sodium atom gives up its outer electron and becomes an Na^+ ion.
The chlorine atom picks up the spare electron and becomes a Cl^- ion.
POP!
Na
Cl
Na
Cl
+
–

Groups 1 and 7 are the Most Likely to Form Ions

1) Ions are charged particles — they can be made from single atoms (e.g. the Cl^- ion) or groups of atoms (e.g. the NO_3^- ion).
2) When atoms lose or gain electrons to form ions, all they're trying to do is get a full outer shell. Atoms like full outer shells — it's atom heaven.
3) Group 1 elements are metals and they lose electrons to form positive ions.
4) Group 7 elements are non-metals. They gain electrons to form negative ions.
5) When an element from Group 1 reacts with an element from Group 7, they form an ionic compound, which you can find out about on page 46. In fact, take a look at them NOW.
6) Molten ionic compounds conduct electricity. This is evidence that they're made up of ions.

Make sure you know ionic compounds inside out and back to front — they're bound to come up in the exam.

Forming ions — seems like the trendy thing to do...

Remember, the +ve and –ve charges we talk about, e.g. Na^+ for sodium, just tell you what type of ion the atom WILL FORM in a chemical reaction. In sodium metal there are only neutral sodium atoms, Na. The Na^+ ions will only appear if the sodium metal reacts with something like water or chlorine.

Ions and Formulas

Once the positive and negative ions have been identified you can work out the formula. Lucky you.

The Charges in an Ionic Compound Add Up to Zero

Different ions have different charges, shown in the table:

Positive Ions		Negative Ions	
Sodium	Na^+	Chloride	Cl^-
Potassium	K^+	Fluoride	Fl^-
Calcium	Ca^{2+}	Bromide	Br^-
Iron(II)	Fe^{2+}	Carbonate	CO_3^{2-}
Iron(III)	Fe^{3+}	Sulfate	SO_4^{2-}

Some metals (like iron, copper and tin) can form ions with different charges. The number in brackets after the name tells you the size of the positive charge on the ion — and luckily for us, this makes the charge really easy to remember. E.g. an iron(II) ion has a charge of 2+, so it's Fe^{2+}.

The main thing to remember is that in compounds the total charge must always add up to zero.

The Easy Ones

If the ions in the compound have the same size charge then it's easy.

EXAMPLE: Find the formula for lithium fluoride.

Find the charges on a lithium ion and a fluoride ion.
A lithium ion is Li^+ and a fluoride ion is F^-.
To balance the total charge you need one lithium ion to every one fluoride ion.
So the formula of lithium fluoride must be: LiF

EXAMPLE: Find the formula for sodium chloride.

Find the charges on a sodium ion and a chloride ion.
A sodium ion is Na^+ and a chloride ion is Cl^-.
To balance the total charge you need one sodium ion to every one chloride ion.
So the formula of sodium chloride must be: $NaCl$

The Harder Ones

If the ions have different size charges, you need to put in some numbers to balance things up.

EXAMPLE: Find the formula for calcium chloride.

Find the charges on a calcium ion and a chloride ion.
A calcium ion is Ca^{2+} and a chloride ion is Cl^-.
To balance the total charge you need two chloride ions to every one calcium ion.
So the formula of calcium chloride must be: $CaCl_2$

EXAMPLE: Find the formula for iron(III) sulfate.

Find the charges on an iron(III) ion and a sulfate ion.
An iron(III) ion is Fe^{3+} and a sulfate ion is SO_4^{2-}.
To balance the total charge you need two iron(III) ions to every three sulfate ions.
So the formula of iron(III) sulfate must be: $Fe_2(SO_4)_3$

You Can Also Work Out the Charges on Ions

If you know the formula of a salt and the charge on one of the ions, you can work out the charge on the other ion:

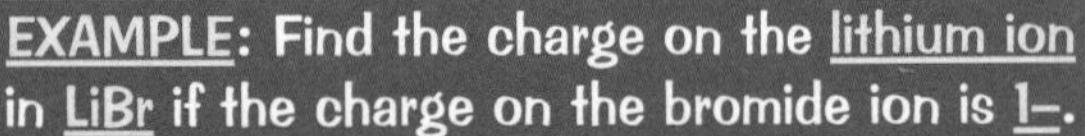

EXAMPLE: Find the charge on the lithium ion in LiBr if the charge on the bromide ion is 1–.

There is one bromide ion (1–) and the charges must balance, so the charge on the lithium ion must be 1+.

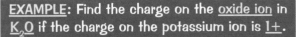

EXAMPLE: Find the charge on the oxide ion in K_2O if the charge on the potassium ion is 1+.

There are two potassium ions (2 × 1+) = 2+. The charges always balance, so the charge on the oxygen ion must be 2–.

Any old ion, any old ion — any, any, any old ion...

* Answers on p.92.

After all those examples, I'm sure you could work out the formula to any ionic compound. And just to test that theory here are a few for you to try: a) magnesium oxide, b) lithium oxide, c) sodium sulfate.*

Revision Summary for Module C4

Okay, if you were just about to turn the page without doing these revision summary questions, then stop. What kind of an attitude is that... Is that really the way you want to live your life... running, playing and having fun... Of course not. That's right. Do the questions. It's for the best all round.

1) What does the nucleus of an atom contain?
2) What is the mass and charge of a neutron?
3)* Balance these equations:
   a) $Na + Cl_2 \rightarrow NaCl$
   b) $K + H_2O \rightarrow KOH + H_2$
4) Write the state symbols for each physical state.
5) A forensic scientist carries out a flame test to identify a metal. The scientist sees a lilac flame. Which metal does this result indicate?
6) Why does each element produce a different line spectrum?
7) What can line spectrums be used for?
8) What feature of atoms determines the order of the periodic table?
9) What are the rows in the periodic table known as?
10) What is significant about the properties of elements in the same group?
11)* Oxygen can be written as $^{16}_{8}O$. How many protons does one atom of oxygen contain?
12) How many electrons can the first shell of any atom hold?
13) Draw the electron configuration of carbon.
14) Which group are the alkali metals?
15) As you go down the group of alkali metals, do they become more or less reactive?
16) Give details of the reactions of the alkali metals with water.
17) Write a balanced equation for the reaction between lithium and chlorine.
18) Describe how the reactivity of the Group 7 elements changes as you go down the group.
19) Atoms of Group 7 elements tend to go round in pairs. What word describes this type of molecule?
20) Describe the appearance of chlorine at room temperature.
21) Describe an experiment that could be used to determine the relative reactivity of the halogens.
22) What does this hazard symbol mean? Give an example of a substance that would need this symbol on its label.
23) What precautions need to be taken when working with Group 7 elements? Why?
24) What is ionic bonding?
25) What kind of atoms like to do ionic bonding? Why is this?
26)* The formula of magnesium bromide is $MgBr_2$. The charge on the bromide ion is 1–. What is the charge on the magnesium ion?
27)* Use the table to help you find the formula for:
   a) iron(II) oxide
   b) iron(III) chloride
   c) calcium oxide
   d) sodium carbonate

Positive ions		Negative ions	
sodium	Na^+	chloride	Cl^-
calcium	Ca^{2+}	oxide	O^{2-}
iron(II)	Fe^{2+}	carbonate	CO_3^{2-}
iron(III)	Fe^{3+}		

* Answers on page 92.

Chemicals in the Atmosphere

Welcome to another section of wonderful Chemistry. Let's kick off with the atmosphere.

Dry Air is a Mixture of Gases

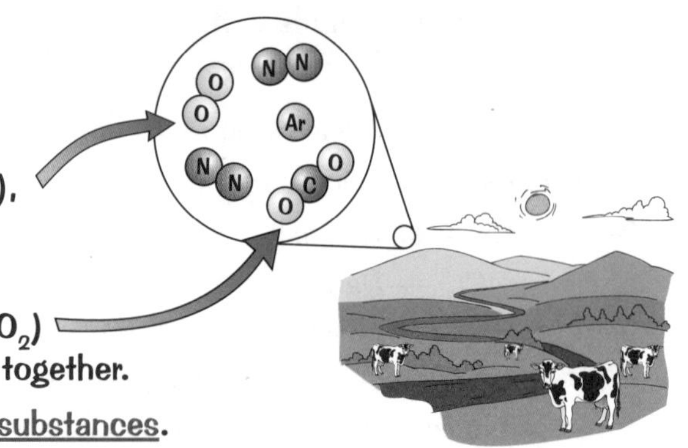

1) The Earth's atmosphere contains many gases.
2) Some of these gases are elements, e.g. oxygen (O_2), nitrogen (N_2) and argon (Ar) — they contain only one type of atom.
3) Other gases are compounds, e.g. carbon dioxide (CO_2) — they contain more than one type of atom joined together.
4) Most of the gases in the atmosphere are molecular substances.

Molecular Substances Have Low Melting and Boiling Points

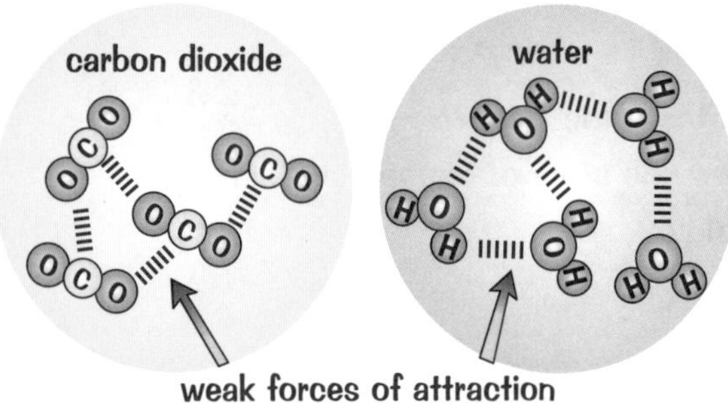

1) Molecular substances usually exist as small molecules, like CO_2 and H_2O.
2) The atoms within the molecules are held together by very strong covalent bonds (see next page).
3) In contrast, the forces of attraction between these molecules are very weak.
4) You only need a little bit of energy to overcome the weak forces between the molecules — so molecular substances have low melting and boiling points.
5) This means that they're usually gases and liquids at room temperature.
6) Pure molecular substances don't conduct electricity, simply because their molecules aren't charged. There are no free electrons or ions.
7) Most non-metal elements and most compounds formed from non-metal elements are molecular.

You Have to be Able to Interpret Data

There's loads of opportunity in this module for examiners to test your ability to interpret data. Here's a nice example on molecular substances.

Example:

Which of the molecular substances in the table is a liquid at room temperature (25 °C)?

	melting point	boiling point
oxygen	-219 °C	-183 °C
nitrogen	-210 °C	-196 °C
bromine	-7 °C	59 °C
argon	-189 °C	-186 °C

There's only one substance that fits the bill here — bromine. It melts (turns to a liquid) at -7 °C and boils (turns to a gas) at 59 °C. So, it'll be a liquid at room temperature. Oxygen, nitrogen and argon will be gases at room temperature.

Stop gassing about it — and get learning...

So, the key things here are those covalent bonds within molecules, and the weaker forces that join the separate molecules together. It's these things that give covalent compounds their properties and make them likely to be liquids and gases at room temperature.

Covalent Bonding

Some elements bond ionically (see page 41), but others form strong covalent bonds.

Covalent Bonds — Sharing Electrons

1) Sometimes atoms make covalent bonds by sharing electrons with other atoms.
2) This way both atoms feel that they have a full outer shell, and that makes them happy.
3) Each covalent bond provides one extra shared electron for each atom.
4) Each atom involved has to make enough covalent bonds to fill up its outer shell.
5) The atoms bond due to the electrostatic attraction between the positive nuclei and the negative electrons shared between them.

E.g. Hydrogen, H_2

Hydrogen needs just one extra electron to fill its outer shell.

So, two hydrogen atoms share their outer electron so that they each have a full shell, and a covalent bond is formed.

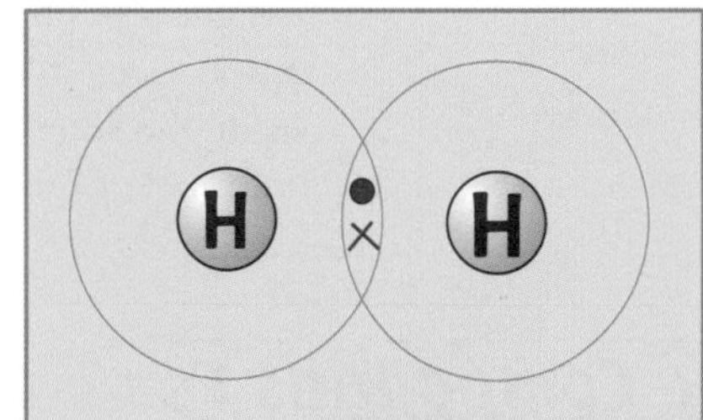

E.g. Carbon Dioxide, CO_2

Carbon needs four more electrons to fill it up.

Oxygen needs two.

So two double covalent bonds are formed.

A double covalent bond has two shared pairs of electrons.

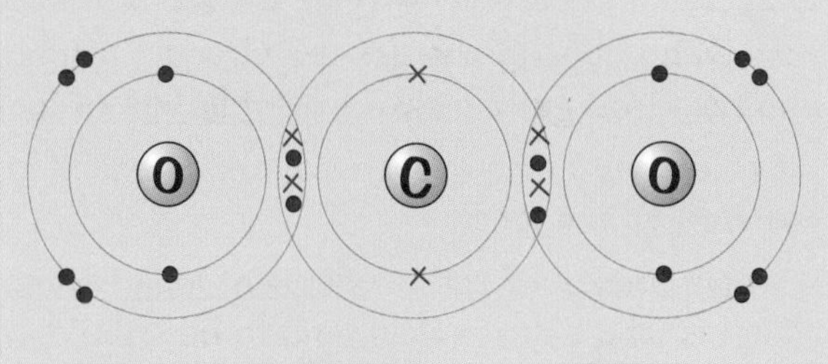

2-D Drawings Don't Always Tell the Whole Story

1) The shape of a molecule is often important, especially with organic (carbon-based) molecules. A simple molecule like methane is often drawn looking flat, but it really has a different shape:

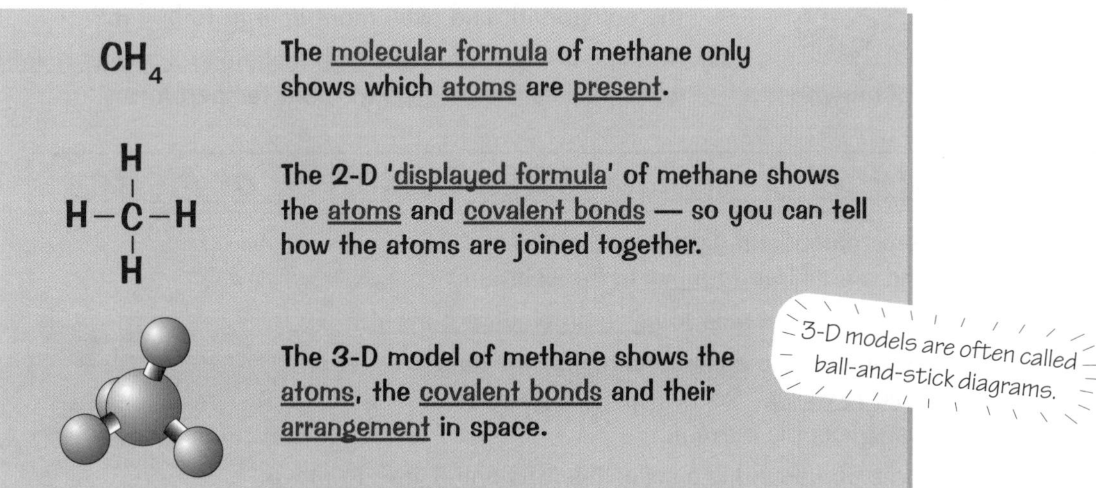

The molecular formula of methane only shows which atoms are present.

The 2-D 'displayed formula' of methane shows the atoms and covalent bonds — so you can tell how the atoms are joined together.

The 3-D model of methane shows the atoms, the covalent bonds and their arrangement in space.

3-D models are often called ball-and-stick diagrams.

2) Computers are often used to produce 3-D models of molecules.

Covalent bonding — it's good to share...

3-D models are especially important when it comes to studying enzymes and some drugs. With enzymes, it's their three-dimensional shape rather than their molecular formula that lets them do their job. There are different types of 3-D model that show you different things about molecules.

Chemicals in the Hydrosphere

The oceans are packed with fish, whales, jellyfish and also plenty of chemicals. Read and learn...

The Earth's Hydrosphere is the Oceans

1) The Earth's hydrosphere consists of all the water in the oceans, seas, lakes, rivers, puddles and so on...
2) It also contains any compounds that are dissolved in the water.
3) Many of these compounds are ionic compounds called salts — that's why sea water is 'salty'.
4) Examples of salts are sodium chloride (NaCl), magnesium chloride ($MgCl_2$) and potassium bromide (KBr).

You need to be able to work out the chemical formulas of salts — here's an example, but flick back to page 42 if you need a recap.

EXAMPLE: Find the formula for magnesium sulfate.

Find the charges on a magnesium ion and a sulfate ion.

A magnesium ion is Mg^{2+} and a sulfate ion is SO_4^{2-}.

To balance the total charge you need one sulfate ion to every one magnesium ion.

So the formula of magnesium sulfate must be: $MgSO_4$

You'll be given a table of charges on ions in the exam.

Solid Ionic Compounds Form Crystals

1) Ionic compounds are made of charged particles called ions.
2) Ions with opposite charges are strongly attracted to one another. You get a massive giant lattice of ions built up.
3) There are very strong chemical bonds called ionic bonds between all the ions.
4) A single crystal of salt is one giant ionic lattice, which is why salt crystals tend to be cuboid in shape.

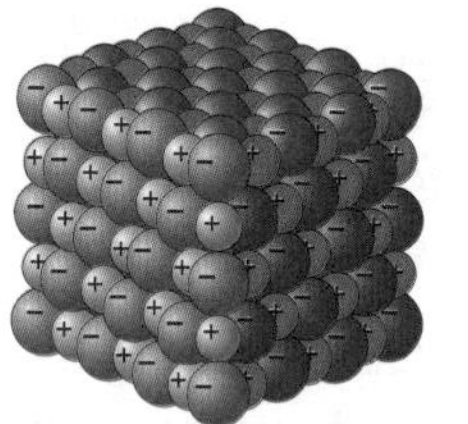

Ionic Compounds Have High Melting and Boiling Points

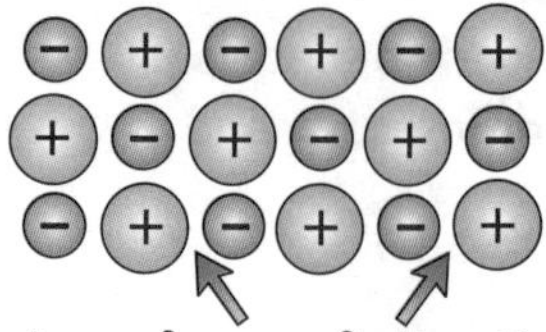

1) The forces of attraction between the ions are very strong.
2) It takes a lot of energy to overcome these forces and melt the compound, and even more energy to boil it.
3) So ionic compounds have high melting and boiling points, which makes them solids at room temperature.

They Conduct Electricity When Dissolved or Molten

1) When an ionic compound dissolves, the ions separate and are all free to move in the solution.
2) This means that they're able to carry an electric current.
3) Similarly, when an ionic compound melts, the ions are again free to move. So — yep, you guessed it — they'll carry electric current.
4) When an ionic compound is a solid, the ions aren't free to move, and so an electrical current can't pass through the substance.

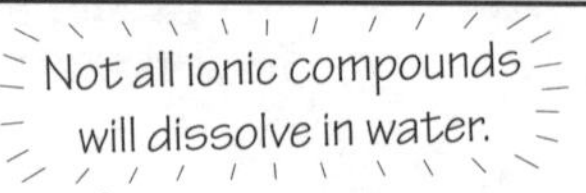

Not all ionic compounds will dissolve in water.

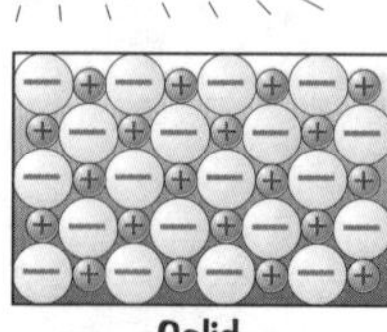

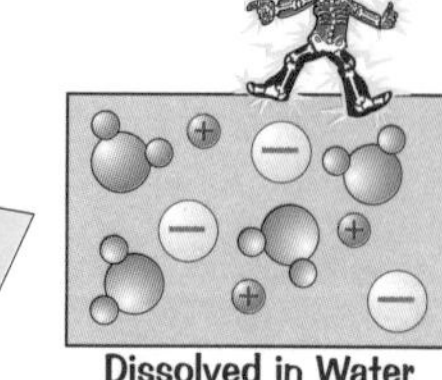

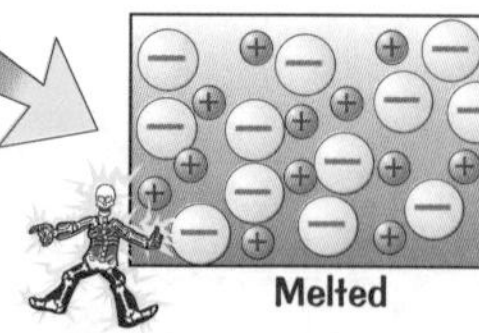

Giant ionic lattices — all over your chips...

Because they conduct electricity when they're dissolved in water, ionic compounds are used to make some types of battery. In the olden days, most batteries had actual liquid in, so they tended to leak all over the place. Now they've come up with a sort of paste that doesn't leak but still conducts. Clever.

Chemicals in the Lithosphere

So, we've done the skies and the seas. Now it's on to the hard stuff (the land, not the work...).

The Earth's Lithosphere is Made Up of a Mixture of Minerals

1) The lithosphere is the Earth's rigid outer layer — the crust and part of the mantle below it.
2) It's made up of a mixture of minerals, often containing silicon, oxygen and aluminium.
3) Most of the silicon and oxygen in the Earth's crust exists as the compound silicon dioxide.

Different types of rock contain different minerals and different elements. For example, limestone contains a lot of calcium, whereas sandstone contains a lot of silicon.

You might have to interpret data on the abundance of elements in rocks. Don't panic — you'll be given all the information you need.

Silicon Dioxide Forms a Giant Covalent Structure

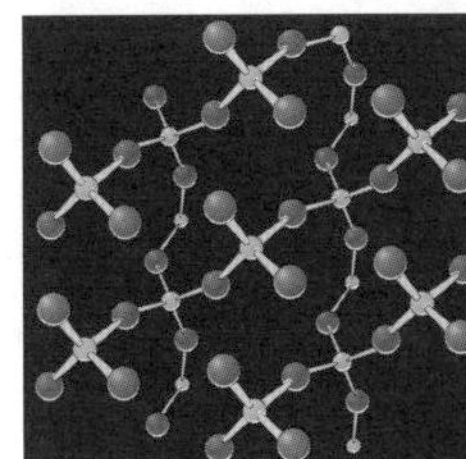

1) Giant covalent structures, like silicon dioxide, contain no charged ions.
2) All the atoms are bonded to each other by strong covalent bonds. This gives it a rigid structure, which makes the substance hard.
3) They have very high melting and boiling points.
4) They don't conduct electricity — not even when molten.
5) They're usually insoluble in water.

Silicon Dioxide

This is what sand is made of. Each grain of sand is one giant structure of silicon and oxygen.
It's also called silica.
Silicon dioxide is the main constituent of sandstone, and it's also found as quartz in granite.

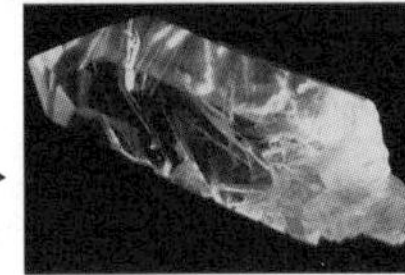

Interpreting data could sneak its way into this section of the exam too...

Example: Diamond has a giant covalent structure. Suggest why it's often used in industrial drill tips.

All the atoms in diamond are bonded to each other by strong covalent bonds, making it an incredibly hard substance — so hard that it can drill through just about any material.

Some Minerals are Very Valuable Gemstones

1) There are a lot of different minerals in the Earth. Some are worth more than others.
2) Some minerals are very rare, which can make them valuable. Gems like diamond, ruby and sapphire are examples of this.
3) Gemstones are very hard. This is all down to their giant covalent structures.
4) They're also pretty and sparkly, which makes them attractive and useful for making jewellery.

Don't forget your minerals — and your vitamins too...

So, all that stuff beneath your feet is packed full of minerals. Some of them are really abundant, but others are quite rare, which can make them pretty valuable. But that doesn't mean you should go off and start digging up your back garden to look for diamonds — right now you need to be learning this.

Chemicals in the Biosphere

The biosphere is anything that's alive, or comes from living things. Again, it's made of chemicals.

Living Things All Share the Same Building Blocks

1) All living things are formed from compounds made up of the same basic elements.
2) The main elements are carbon, hydrogen, oxygen and nitrogen, along with small amounts of phosphorus and sulfur.
3) These elements make up molecules vital for life, such as carbohydrates, proteins, fats and DNA.

It's possible to recognise molecules by the elements they contain.

1) DNA always contains phosphorus and nitrogen.
2) Proteins always contain nitrogen and may also contain sulfur.
3) Fats and carbohydrates only contain carbon, hydrogen and oxygen... you can tell them apart as fats contain a greater percentage of carbon.

This is another 'interpret data' topic, so make sure you know what makes up each molecule.

4) Don't forget, carbohydrates, proteins and DNA are all molecular — there are no ions involved.

You Can Write Formulas by Counting Elements

For example, the carbohydrate glucose looks like this:

```
          CH2OH
            |
            C — O
      H   / |     \   H
       \ /  H      \ /
        C           C
       / \ OH  H   / \
     OH   \ |  |  /   OH
            C — C
            |   |
            H   OH
```

The diagram shows that it only contains carbon, hydrogen and oxygen atoms.

There are 6 carbon atoms, 12 hydrogen atoms and 6 oxygen atoms, so you can work out that glucose's formula is $C_6H_{12}O_6$.

Flow Charts Show Changes Between Spheres

1) Elements are constantly moving between the atmosphere, biosphere, hydrosphere and lithosphere.
2) Flow charts can be used to summarise chemical changes between the spheres.

   1) Arrows show the direction of change.
   2) Boxes represent the various stages.

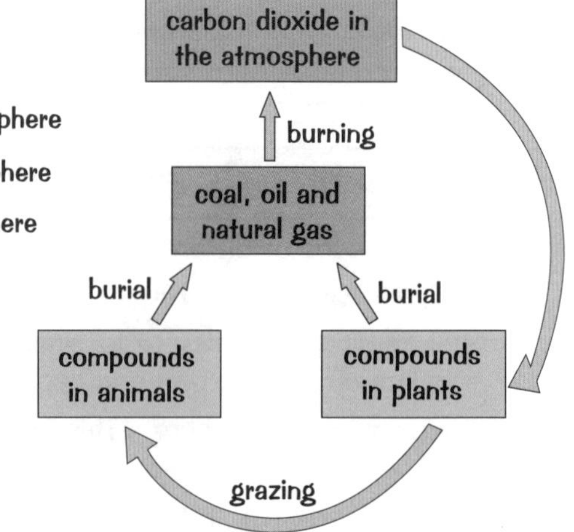

The diagram shows part of the carbon cycle.

Carbon in the atmosphere (carbon dioxide) is captured by green plants and converted into other carbon compounds, like carbohydrates. Animals graze on plants, taking in the carbon compounds they contain. The plants and animals die and some are buried. Over millions of years these form coal and oil, which, when burnt, will return the carbon dioxide to the atmosphere.

ONE carbon atom — ah, ah, ah, ah...

Examiners love to ask you to interpret information like flow charts and molecule data. Don't panic if it's something you don't know about. You won't be asked questions about specific facts — you just need to be able to understand what the data means. So take a deep breath and use your common sense.

Metals from Minerals

Ores Contain Enough Metal to Make Extraction Worthwhile

1) Rocks are made of minerals. Minerals are just solid elements and compounds.
2) Metal ores are rocks that contain varying amounts of minerals from which metals can be extracted.
3) In many cases the ore is an oxide of the metal. Here are a few examples of ores:

> a) A type of iron ore is called haematite. This is iron(III) oxide (Fe_2O_3).
> b) A type of copper ore is called chalcopyrite. This is copper iron sulfide ($CuFeS_2$).

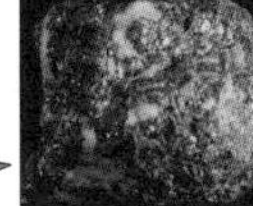
Chalcopyrite

4) For some metals, large amounts of ore need to be mined just to obtain small percentages of valuable minerals. A good example of this is copper mining — copper ores typically contain about 1% copper.

You may be asked to calculate the mass of a metal that can be extracted from a certain mass of mineral given its formula (see below) or an equation (see p.60).

> Example: How much copper can be extracted from 800 g of copper oxide (CuO)?
>
> Step 1: Calculate the proportion of copper in copper oxide $= \frac{A_r \times \text{no. of atoms}}{M_r} = \frac{63.5 \times 1}{79.5} = 0.79874$
>
> Step 2: Multiply your answer by the mass of copper oxide $= 0.79874 \times 800 = 639$ g
>
> See p.59 for more about A_r and M_r.

More Reactive Metals are Harder to Get

1) A few unreactive metals like gold are found in the Earth as the metal itself, rather than as a compound.
2) But most metals need to be extracted from their ores using a chemical reaction.
3) More reactive metals, like sodium, are harder to extract — that's why it took longer to discover them.

Some Metals can be Extracted by Reduction with Carbon

1) Electrolysis (splitting with electricity) is one way of extracting a metal from its ore (see next page). The other common way is chemical reduction using carbon or carbon monoxide.
2) When an ore is reduced, oxygen is removed from it, e.g.

$Fe_2O_3(s)$	+	$3CO(g)$	→	$2Fe(s)$	+	$3CO_2(g)$
iron(III) oxide	+	carbon monoxide	→	iron	+	carbon dioxide

3) When a metal oxide loses its oxygen it is **REDUCED**. The carbon gains the oxygen and is **OXIDISED**.

The position of a metal in the reactivity series determines whether it can be extracted by reduction with carbon or carbon monoxide.

a) Metals higher than carbon in the reactivity series have to be extracted using electrolysis, which is expensive.

b) Metals below carbon in the reactivity series can be extracted by reduction using carbon.
This is because carbon can only take the oxygen away from metals which are less reactive than carbon itself is.

The Reactivity Series		
Sodium	Na	more
Calcium	Ca	reactive
Magnesium	Mg	
Aluminium	Al	
CARBON	C	
Zinc	Zn	
Iron	Fe	
Tin	Sn	less
Copper	Cu	reactive

Miners — they always have to stick their ore in...

Extracting metals isn't cheap. You have to pay for special equipment, energy and labour. Then there's the cost of getting the ore to the extraction plant. If there's a choice of extraction methods, a company always picks the cheapest, unless there's a good reason not to. They're not extracting it for fun.

Electrolysis

Electrolysis is a useful way of extracting reactive metals from their ores.

Electrolysis Means 'Splitting Up with Electricity'

1) Electrolysis is the decomposition (breaking down) of a substance using electricity.
2) It needs a liquid to conduct the electricity — called the electrolyte. Electrolytes are usually free ions dissolved in water (e.g. dissolved salts) or molten ionic compounds.
3) It's the free ions that conduct the electricity and allow the whole thing to work.
4) For an electrical circuit to be complete, there's got to be a flow of electrons. In electrolysis, electrons are taken away from ions at the positive electrode and given to other ions at the negative electrode. As ions gain or lose electrons they become atoms or molecules.

NaCl dissolved

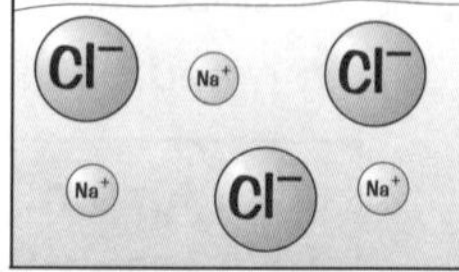

Molten NaCl

Electrolysis Removes Aluminium from Its Ore

1) The main ore of aluminium is bauxite, which contains aluminium oxide, Al_2O_3.
2) Molten aluminium oxide contains free ions — so it'll conduct electricity.
3) The positive Al^{3+} ions are attracted to the negative electrode where they each pick up three electrons and "zup", they turn into neutral aluminium atoms. These then sink to the bottom.
4) The negative O^{2-} ions are attracted to the positive electrode where they each lose two electrons. The neutral oxygen atoms will then combine to form O_2 molecules.

Metals form positive ions, so they're attracted to the negative electrode.

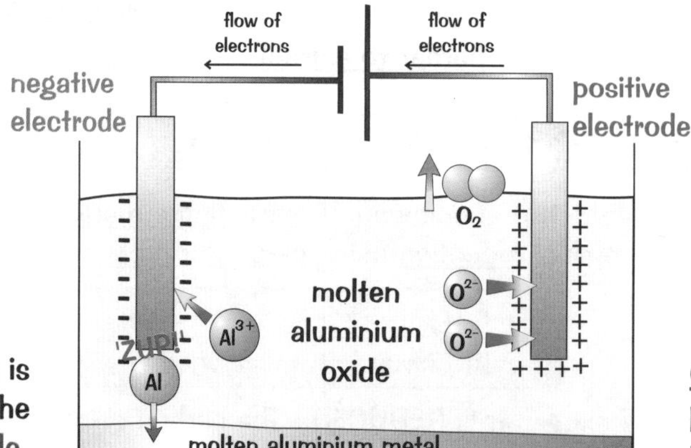

Non-metals form negative ions, so they're attracted to the positive electrode.

Aluminium is produced at the negative electrode.

Oxygen is produced at the positive electrode.

At the Negative Electrode:

$$Al^{3+} + 3e^- \rightarrow Al$$

Reduction — a gain of electrons

At the Positive Electrode:

$$2O^{2-} \rightarrow O_2 + 4e^-$$

Oxidation — a loss of electrons

So, the complete equation for the decomposition of aluminium oxide is:

aluminium oxide → aluminium + oxygen

$$2Al_2O_{3(l)} \rightarrow 4Al_{(l)} + 3O_{2(g)}$$

Always remember state symbols when writing symbol equations.

Faster shopping at Tesco — use Electrolleys...

Electrolysis ain't cheap — it takes a lot of electricity, which costs money. It's the only way of extracting some metals from their ores though, so it's worth it. This isn't such a bad page to learn — try writing a mini-essay about it. Don't forget to have a go at drawing the diagram from memory too.

Metals

Who'd have thought you'd find metals lurking about in rocks...
Now you've seen how to extract them, it's time to learn all about metals and their properties, yay!

Metal Properties are All Due to the Sea of Free Electrons

1) Metals consist of a giant structure.
2) Metallic bonds involve the all-important 'free electrons', which produce all the properties of metals.
3) These free electrons come from the outer shell of every metal atom in the structure.
4) The positively charged metal ions are held together by these electrons.

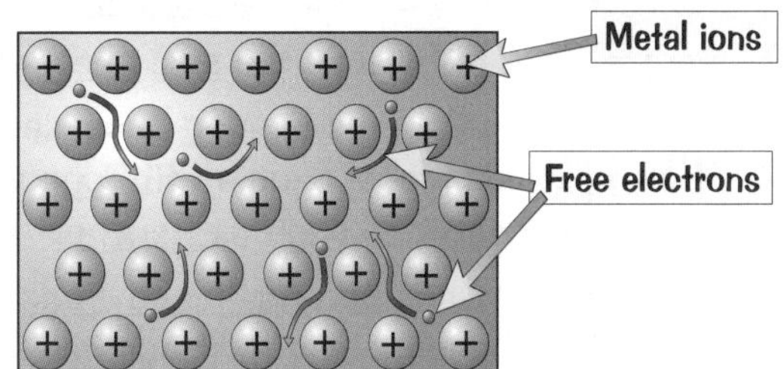

1) They're Good Conductors of Heat and Electricity

The free electrons carry both heat and electrical current through the material, so metals are good conductors of heat and electricity.

> Metals are ideal if you want to make something that heat needs to travel through, like a saucepan base.
> Their electrical conductivity makes them great for making things like electrical wires.

2) Most Metals are Strong and Malleable

Metals have a high tensile strength — in other words they're strong and hard to break.

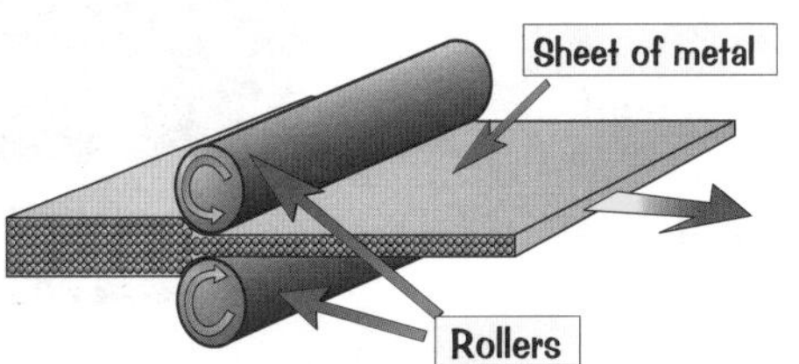

The layers of atoms in a metal can slide over each other, making metals malleable — they can be hammered or rolled into flat sheets.

> Metals' strength and 'bendability' makes them handy for making into things like bridges and car bodies.

3) They Generally Have High Melting and Boiling Points

Metallic bonds are very strong, so it takes a lot of energy to break them — you have to get the metal pretty hot to melt it (except for mercury, which is a bit weird), e.g. copper melts at 1085 °C and tungsten melts at 3422 °C.

> Metals' high melting and boiling points make them handy — you don't want your saucepan to melt when you're cooking, or bridges to melt in hot weather.

Someone robbed your metal? — call a copper...

The skin of the Statue of Liberty is made of copper — about 80 tonnes of it in fact. Its surface reacts with gases in the air to form copper carbonate — which is why it's that pretty shade of green. It was a present from France to the United States — I wonder if they found any wrapping paper big enough?

Environmental Impact

Metals are definitely a big part of modern life. Once they're finished with, it's far better to recycle them than to dig up more ore and extract fresh metal.

Ores are Finite Resources

1) This means that there's a limited amount of them — eventually, they'll run out.
2) People have to balance the social, economic and environmental effects of mining the ores.
3) So, mining metal ores is good because useful products can be made. It also provides local people with jobs and brings money into the area. This means services such as transport and health can be improved.
4) But mining ores is bad for the environment as it uses loads of energy, scars the landscape and destroys habitats. Also, noise, dust and pollution are caused by an increase in traffic.
5) Deep mine shafts can also be dangerous for a long time after the mine has been abandoned.

Recycling Metals is Important

1) Mining and extracting metals takes lots of energy, most of which comes from burning fossil fuels.
2) Fossil fuels are running out so it's important to conserve them. Not only this, but burning them contributes to acid rain, global dimming and probably climate change.
3) Recycling metals only uses a small fraction of the energy needed to mine and extract new metal. E.g. recycling copper only takes 15% of the energy that's needed to mine and extract new copper.
4) Energy doesn't come cheap, so recycling saves money too.
5) As there's a finite amount of each metal in the Earth, recycling conserves these resources.
6) Recycling metal cuts down on the amount of rubbish that gets sent to landfill. Landfill takes up space and pollutes the surroundings. If all the aluminium cans in the UK were recycled, there'd be 14 million fewer dustbins of waste each year.

For example...

1) If you didn't recycle, say, aluminium, you'd have to mine more aluminium ore — 4 tonnes for every 1 tonne of aluminium you need. But mining makes a mess of the landscape (and these mines are often in rainforests). The ore then needs to be transported, and the aluminium extracted (which uses loads of electricity). And don't forget the cost of sending your used aluminium to landfill.
2) So it's a complex calculation, but for every 1 kg of aluminium cans you recycle, you save:
   - 95% or so of the energy needed to mine and extract 'fresh' aluminium,
   - 4 kg of aluminium ore,
   - a lot of waste.

In fact, aluminium's about the most cost-effective metal to recycle.

Recycling — do the Tour de France twice...

Recycling metal saves natural resources and money and reduces environmental problems. It's great. There's no limit to the number of times metals like aluminium, copper and steel can be recycled. So your humble little drink can may one day form part of a powerful robot who takes over the galaxy.

Revision Summary for Module C5

Here are some questions for you to get your teeth into. Have a go at them. If there are any you can't do, go back to the section and do a bit more learning, then try again. It's not fun, but it's the best way to make sure you know everything. Hop to it.

1) Name a gas in dry air that is a compound. Give the formula of this gas.
2) Explain why most molecular substances are gases.
3) Why don't molecular compounds conduct electricity?
4) What is covalent bonding?
5)* Ethane can be represented by the molecular formula C_2H_6. Draw a 2-D diagram of its structure.
6) What is the Earth's hydrosphere?
7) What makes sea water 'salty'?
8) Why do solid ionic compounds form crystals?
9) Do solid ionic compounds have low or high boiling points? Explain why.
10) Why can ionic compounds conduct electricity when dissolved in water but not when they're solid?
11) What is the Earth's lithosphere?
12) Name three abundant elements in the Earth's lithosphere.
13) What type of structure does silicon dioxide have?
14) Give two chemical properties of silicon dioxide.
15) Why are some minerals considered to be valuable gemstones?
16)* From the table, which molecule is more likely to be a fat, and which is more likely to be a carbohydrate?

element	molecule A	molecule B
carbon	40%	75%
hydrogen	6%	13%
oxygen	54%	12%

17) In a flowchart, what do: a) the arrows represent? b) the boxes represent?
18) What is an ore?
19)* Calculate how much aluminium can be extracted from 400 g of aluminium oxide.
20) Name a metal that can be extracted by heating its oxide with carbon.
21) Why can't some metal oxides be reduced using carbon?
22) What is electrolysis?
23) Why can a molten ionic crystal act as an electrolyte?
24) During electrolysis, do metals form at the negative electrode or at the positive electrode?
25) Describe what happens at the electrodes during the electrolysis of molten aluminium oxide.
26) What do the terms 'reduced' and 'oxidised' mean?
27) Why are metals able to conduct electricity?
28) Why have metals got high melting and boiling points?
29) Give one impact on the environment for each of the following:
    a) extracting metals,
    b) disposing of metals.
30) Give two reasons why it's a good idea to recycle metals.

* Answers on page 92.

Industrial Chemical Synthesis

Chemical synthesis is the term used by those in the know to describe the process of making complex chemical compounds from simpler ones. That's what this section is all about — some of the methods chemists use to make and purify chemicals, the calculations they carry out and how they control the reactions to make the maximum amount of product and ultimately the maximum amount of money.

The Chemical Industry Makes Useful Products

Chemicals aren't just found in the laboratory. Most of the products you come across in your day-to-day life will have been carefully researched, formulated and tested by chemists. Here are a few examples...

1) Food additives — the chemical industry produces additives like preservatives, colourings and flavourings for food producers.
2) Cleaning and decorating products — things like paints contain loads of different pigments and dyes, all of which have been developed by chemists. Cleaning products like bleach, oven cleaner and washing-up liquid will all have been developed by chemists.

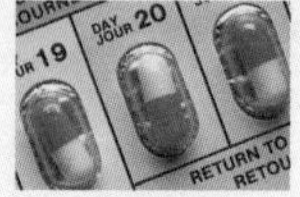

3) Drugs — the pharmaceutical industry is huge (see below). Whenever you have a headache or an upset tummy the drugs you take will have gone through loads of development and testing before you get to use them.
4) Fertilisers — we use about a million tonnes of fertiliser every year. Amongst other things fertilisers contain loads of ammonia, all of which has to be produced by the chemical industry.

As well as figuring out how to make chemicals, chemists must also figure out how to make them in the way that produces the highest yield (p.61) — they do this by controlling the rate of the reaction (p.64). They must also think about the environment, choosing processes with a low impact.

The Chemical Industry is Huge

It's absolutely massive, in terms of both the amount of chemicals it produces and the money it generates.

Scale — chemicals can be produced on a large or small scale.

1) Some chemicals are produced on a massive scale — for example, over 150 million tonnes of sulfuric acid are produced around the world every year.
2) Sulfuric acid has loads of different uses, for example in car batteries and fertiliser production.
3) Other chemicals, e.g. pharmaceuticals, are produced on a smaller scale, but this doesn't make them any less important — we just need less of them.

Sectors — there are loads of different sectors within the chemical industry.

1) In the UK, the chemical industry makes up a significant chunk of the economy.
2) In the UK alone, there are over 200 000 people employed in the chemical industry.
3) Some chemicals are sold directly to consumers, while others are sold to other industries as raw materials for other products.
4) The pharmaceutical industry has the largest share of the industry.

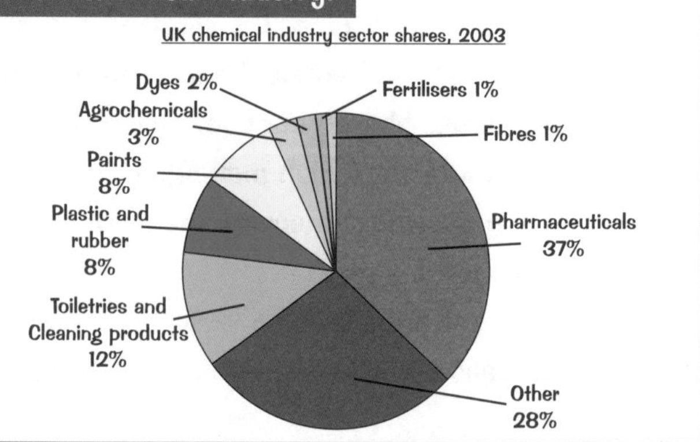

You'd need a big fish to make all those chemicals on a scale...

You don't need to remember all the figures on the pie chart but they may ask you to interpret a similar one in the exam. Don't worry though — everything you'll need to answer the questions will be there.

Acids and Alkalis

You'll find acids and alkalis at home, in the lab and in industry — they're an important set of chemicals.

Substances can be Acidic, Alkaline or Neutral

There's a sliding scale from very strong acid (pH 0) to very strong alkali (pH 14).

These are the colours you get when you add universal indicator to an acid or an alkali.

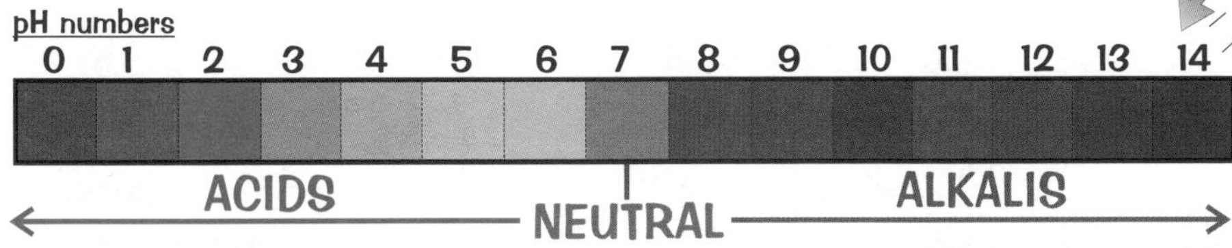

Pure acidic compounds can be solids (e.g. citric acid, which is used as a food additive, and tartaric acid), liquids (e.g. sulfuric acid, nitric acid and ethanoic acid, which is the acid in vinegar) or gases (e.g. hydrogen chloride).

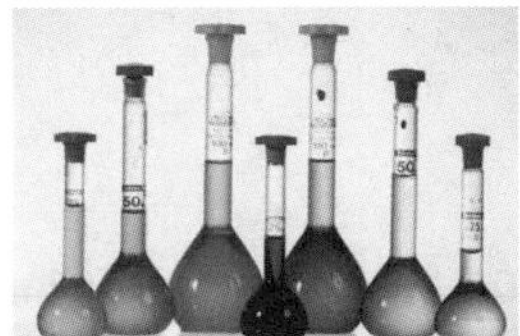

Common alkalis include sodium hydroxide (which is used to make cleaning products like bleach), potassium hydroxide (used in alkaline batteries) and calcium hydroxide (which can be used to neutralise acidic soils).

Indicators and pH Meters can be Used to Determine pH

1) Indicators contain a dye that changes colour depending on whether it's above or below a certain pH.
2) Universal indicator is a very useful combination of dyes, which gives the colours shown above. It's useful for estimating the pH of a solution.
3) pH meters can also be used to measure the pH of a substance. These usually consist of a probe, which is dipped into the substance, and a meter, which gives a reading of the pH.
4) pH meters are more accurate than indicators.

Neutralisation Reactions Between Acids and Alkalis Make Salts

An ACID is a substance with a pH of less than 7.
Acidic compounds form aqueous hydrogen ions, $H^+(aq)$, in water.
An ALKALI is a substance with a pH of greater than 7.
Alkaline compounds form aqueous hydroxide ions, $OH^-(aq)$, in water.

An acid and an alkali react together to form a salt and water. The products of the reaction aren't acidic or alkaline — they're neutral. So it's called a neutralisation reaction.
The general equation is the same for any neutralisation reaction, so make sure you learn it:

$$\text{acid} + \text{alkali} \rightarrow \text{salt} + \text{water}$$

Neutralisation can also be seen in terms of H^+ and OH^- ions. The hydrogen ions from the acid react with the hydroxide ions from the alkali to make water (which is neutral).

$$H^+_{(aq)} + OH^-_{(aq)} \rightarrow H_2O_{(l)}$$

All my indicators are orange...

There's no getting away from acids and alkalis in Chemistry, or even in real life. They're everywhere — acids are found in loads of foods, either naturally like in fruit, or as flavourings and preservatives, whilst alkalis (particularly sodium hydroxide) are used to help make all sorts of things from soaps to ceramics.

Acids Reacting with Metals

Not only do you need to know about how acids react with alkalis but also how they react with metals.

Acid + Metal → Salt + Hydrogen

That's written big 'cos it's kinda worth remembering. Here's the typical experiment:

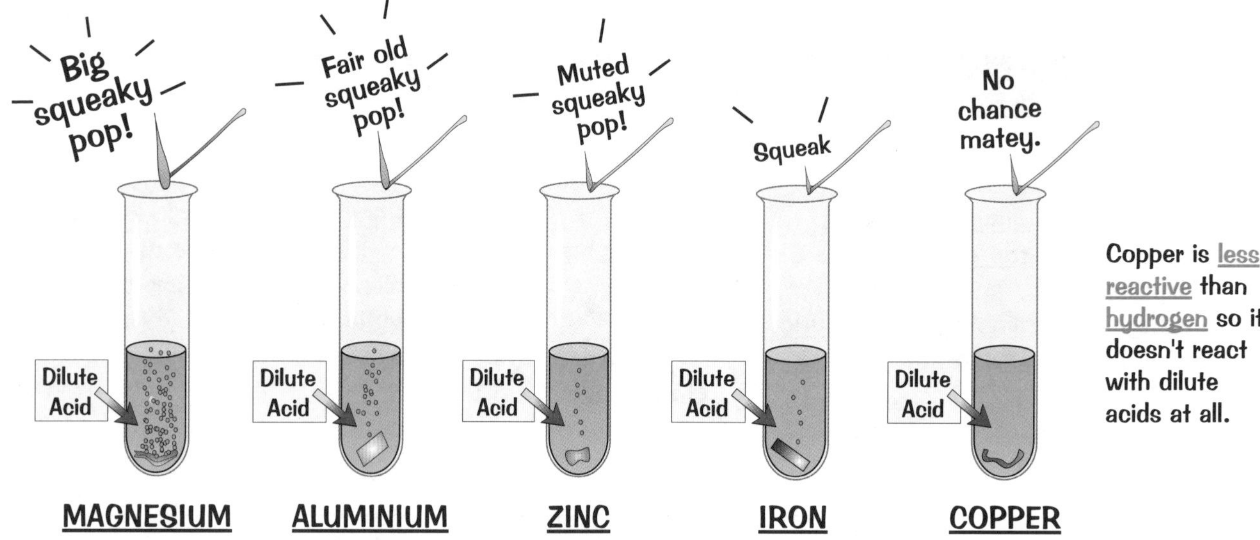

Copper is less reactive than hydrogen so it doesn't react with dilute acids at all.

1) The more reactive the metal, the faster the reaction will go — very reactive metals (e.g. sodium) react explosively.
2) Copper does not react with dilute acids at all — because it's less reactive than hydrogen.
3) The speed of reaction is indicated by the rate at which the bubbles of hydrogen are given off.
4) The hydrogen is confirmed by the burning splint test giving the notorious 'squeaky pop'.
5) The name of the salt produced depends on which metal is used, and which acid is used:

Hydrochloric Acid Will Always Produce Chloride Salts:

$2HCl(aq) + Mg(s) \rightarrow MgCl_2(aq) + H_2(g)$ (Magnesium chloride)

$6HCl(aq) + 2Al(s) \rightarrow 2AlCl_3(aq) + 3H_2(g)$ (Aluminium chloride)

$2HCl(aq) + Zn(s) \rightarrow ZnCl_2(aq) + H_2(g)$ (Zinc chloride)

You need to be able to write balanced symbol equations — see p.34 for more.

Sulfuric Acid Will Always Produce Sulfate Salts:

$H_2SO_4(aq) + Mg(s) \rightarrow MgSO_4(aq) + H_2(g)$ (Magnesium sulfate)

$3H_2SO_4(aq) + 2Al(s) \rightarrow Al_2(SO_4)_3(aq) + 3H_2(g)$ (Aluminium sulfate)

$H_2SO_4(aq) + Zn(s) \rightarrow ZnSO_4(aq) + H_2(g)$ (Zinc sulfate)

Remember to include state symbols. Chloride and sulfate salts are generally soluble in water so they get an aqueous (aq) state symbol (the main exceptions are lead chloride, lead sulfate and silver chloride, which are insoluble).

Nitric Acid Produces Nitrate Salts When NEUTRALISED, But...

Nitric acid (HNO_3) reacts fine with alkalis, to produce nitrates, but it can play silly devils with metals and produce nitrogen oxides instead, so we'll ignore it here. Chemistry's a real messy subject sometimes.

Nitric acid, tut — there's always one...

Some of these reactions are really useful, and some are just for fun (who said Chemistry was dull). Try writing equations for different combinations of acids and metals. Balance them. Cover the page and scribble all the equations down. If you make any mistakes just try again...

Oxides, Hydroxides and Carbonates

Here's more stuff on neutralisation reactions — mixing acids with various alkalis and carbonates.

Metal Oxides and Metal Hydroxides React with Acids

All metal oxides and hydroxides react with acids to form a salt and water.

Acid + Metal Oxide → Salt + Water

Acid + Metal Hydroxide → Salt + Water

These are neutralisation reactions.

The Combination of Metal and Acid Decides the Salt

Here are a couple of examples of metal oxides reacting with acids:

Here the copper ion is Cu^{2+}, so it needs two Cl^- ions.

hydrochloric acid	+	copper oxide	→	copper chloride	+	water
$2HCl(aq)$	+	$CuO(s)$	→	$CuCl_2(aq)$	+	$H_2O(l)$
sulfuric acid	+	zinc oxide	→	zinc sulfate	+	water
$H_2SO_4(aq)$	+	$ZnO(s)$	→	$ZnSO_4(aq)$	+	$H_2O(l)$

And here are a couple of examples of metal hydroxides reacting with acids:

hydrochloric acid	+	sodium hydroxide	→	sodium chloride	+	water
$HCl(aq)$	+	$NaOH(aq)$	→	$NaCl(aq)$	+	$H_2O(l)$
sulfuric acid	+	calcium hydroxide	→	calcium sulfate	+	water
$H_2SO_4(aq)$	+	$Ca(OH)_2(aq)$	→	$CaSO_4(aq)$	+	$2H_2O(l)$

The sulfate ion is SO_4^{2-}, so it needs two H^+ ions.

The calcium ion is Ca^{2+}, so it needs two OH^- ions.

See p.42 for more on finding formulas.

Metal Carbonates Give Salt + Water + Carbon Dioxide

More gripping reactions involving acids. At least there are some bubbles involved here.

Acid + Metal Carbonate → Salt + Water + Carbon Dioxide

The reaction is the same as any other neutralisation reaction EXCEPT that carbonates give off carbon dioxide as well. Practise writing the following equations out from memory — it'll do you no harm at all.

hydrochloric acid + sodium carbonate → sodium chloride + water + carbon dioxide

$$2HCl(aq) + Na_2CO_3(s) \rightarrow 2NaCl(aq) + H_2O(l) + CO_2(g)$$

Here's another example. (Notice how the equation's quite similar.)

hydrochloric acid + calcium carbonate → calcium chloride + water + carbon dioxide

$$2HCl(aq) + CaCO_3(s) \rightarrow CaCl_2(aq) + H_2O(l) + CO_2(g)$$

Someone threw some NaCl at me — I said, "Hey that's a salt"...

The acid + carbonate reaction is one you might have to do at home. If you live in a hard water area, you'll get insoluble $MgCO_3$ and $CaCO_3$ 'furring up' your kettle. You can get rid of this with 'descaler', which is dilute acid (often citric acid) — this reacts with the insoluble carbonates to make soluble salts.

Synthesising Compounds

As I'm sure you know, the chemical industry is really important. Without it we'd be without loads of everyday chemicals. When it comes to making these chemicals it's not just a case of throwing everything into a bucket — oh no, there are quite a few stages to the process — seven, to be precise.

There are Seven Stages Involved in Chemical Synthesis

1 CHOOSING THE REACTION

Chemists need to choose the reaction (or series of reactions) to make the product. For example:

- neutralisation (see p.55) — an acid and an alkali react to produce a salt.
- thermal decomposition — heat is used to break up a compound into simpler substances.
- precipitation — an insoluble solid is formed when two solutions are mixed.

2 RISK ASSESSMENT

This is an assessment of anything in the process that could cause injury (see p.40)

It involves:
- identifying hazards
- assessing who might be harmed
- deciding what action can be taken to reduce the risk.

3 CALCULATING THE QUANTITIES OF REACTANTS

This includes a lot of maths and a balanced symbol equation (p.34). Using the equation chemists can calculate how much of each reactant is needed to produce a certain amount of product. This is particularly important in industry because you need to know how much of each raw material is needed so there's no waste — waste costs money.

4 CHOOSING THE APPARATUS AND CONDITIONS

The reaction needs to be carried out using suitable apparatus and in the right conditions. The apparatus needs to be the correct size (for the amount of product and reactants) and strength (for the type of reaction being carried out, e.g. if it is explosive or gives out a lot of heat). Chemists need to decide what temperature the reaction should be carried out at, what concentrations of reactants should be used, and whether or not to use a catalyst (see p.64).

5 ISOLATING THE PRODUCT

After the reaction is finished the products may need to be separated from the reaction mixture. This could involve evaporation (if the product is dissolved in the reaction mixture), filtration (if the product is an insoluble solid) and drying (to remove any water) see p.61.

6 PURIFICATION

Isolating the product and purification go together like peas and carrots. As you're isolating the product you're also helping to purify it. Crystallisation can be useful in the purification process.

7 MEASURING YIELD AND PURITY

The yield tells you about the overall success of the process. It compares what you think you should get with what you get in practice (see p.61). The purity of the chemical also needs to be measured (p.63).

It's just like Snow White — but with chemical synthesis steps...

It's important that none of these stages are missed out. It'd be pointless if you weren't able to separate the product from the reaction mixture and even worse if the process caused injury or death. Be safe.

Relative Formula Mass

One of the most important stages in chemical synthesis is deciding the mass of reactants needed. Careful calculations mean less waste, and less waste means more profit. But to do all that you'll need to understand relative atomic mass and relative formula mass. It's not as bad as is sounds...

Relative Atomic Mass, A_r — Easy Peasy

1) This is just a way of saying how heavy different atoms are compared with the mass of an atom of carbon-12. So carbon-12 has an A_r of exactly 12.
2) You can work out an element's the relative atomic mass by looking at the periodic table.
3) In the periodic table, the elements all have two numbers. The bigger one is the relative atomic mass — for more on this see p.36.

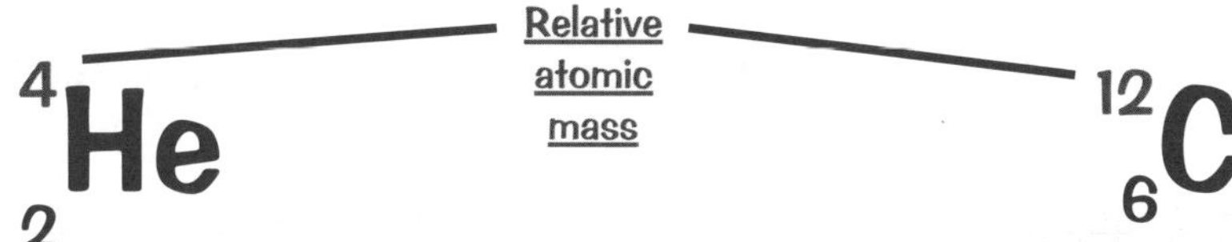

Helium has A_r = 4. Carbon has A_r = 12. Chlorine has A_r = 35.5.

Relative Formula Mass, M_r — Also Easy Peasy

If you have a compound like $MgCl_2$ then it has a relative formula mass, M_r, which is just all the relative atomic masses added together.

For $MgCl_2$ it would be:

$MgCl_2$

24 + (35.5 × 2) = 95

So the M_r for $MgCl_2$ is simply 95.

You can easily get the A_r for any element from the periodic table (see p.36).
I'll tell you what, since it's nearly Christmas I'll run through another example for you:

Compounds with Brackets in...

Find the relative formula mass for magnesium hydroxide, $Mg(OH)_2$.

ANSWER: The small number 2 after the bracket in the formula $Mg(OH)_2$ means that there's two of everything inside the brackets. But that doesn't make the question any harder really.

The brackets in the sum are in the same place as the brackets in the chemical formula.

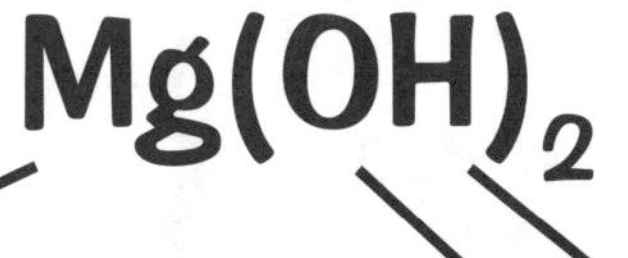

So the relative formula mass for $Mg(OH)_2$ is 58.

(1 × 24) + [(16 + 1) × 2] = 58

And that's all it is. A big fancy name like relative formula mass and all it means is "add up all the relative atomic masses". What a swizz, eh? You'd have thought it'd be something a bit juicier than that, wouldn't you. Still, that's life — it's all a big disappointment in the end. Sigh.

Numbers? — and you thought you were doing chemistry...

Learn the definitions of relative atomic mass and relative formula mass, then have a go at these:

1) Use the periodic table to find the relative atomic mass of these elements: Cu, K, Kr, Cl
2) Also find the relative formula mass of these compounds: NaOH, HNO_3, KCl, $CaCO_3$

Answers on page 92.

Calculating Masses in Reactions

Once you've mastered relative formula masses you can calculate the masses in reactions.

The Three Important Steps — Not to be Missed...

(Miss one out and it'll all go horribly wrong, believe me.)

1) Write out the balanced equation.
2) Work out M_r — just for the two bits you want.
3) Apply the rule: Divide to get one, then multiply to get all.
   (But you have to apply this first to the substance they give information about, and then the other one!)

Don't worry — these steps should all make sense when you look at the example below.

Example: What mass of magnesium is needed to produce 100 g of magnesium oxide?

Answer:

See page 34 for how to write a balanced equation.

1) Write out the balanced equation: $2Mg + O_2 \rightarrow 2MgO$

2) Work out the relative formula masses: $2 \times 24 \rightarrow 2 \times (24 + 16)$
   (don't do the oxygen — you don't need it) $48 \rightarrow 80$

3) Apply the rule: Divide to get one, then multiply to get all.
   The two numbers, 48 and 80, tell us that 48 g of Mg react to give 80 g of MgO.
   Here's the tricky bit. You've now got to be able to write this down:

48 g of Mg ......reacts to give.....80 g of MgO
? g of Mg ......reacts to give.....1 g of MgO
? g of Mg .....reacts to give......100 g of MgO

The big clue is that in the question they've said that 100 g of magnesium oxide is produced, i.e. they've told us how much magnesium oxide to have, and that's how you know to fill in the right-hand side of the box first, because:

We'll first need to ÷ by 80 to get 1 g of MgO
and then need to × by 100 to get 100 g of MgO.

Then you can work out the numbers on the other side (shown in red below) by realising that you must divide both sides by 80 and then multiply both sides by 100. It's tricky.

÷80	48 g of Mg	80 g of MgO	÷80
	0.6 g of Mg	1 g of MgO	
×100	60 g of Mg	100 g of MgO	×100

The mass of product is called the yield of a reaction. Masses you calculate in this way are called **THEORETICAL YIELDS**. In practice you never get 100% of the yield, so the amount of product will be less than calculated (see p. 61).

This tells us that 60 g of magnesium is needed to produce 100 g of magnesium oxide. If the question had said, "What mass of magnesium oxide is produced when 60 g of magnesium is burned in air", you'd fill in the Mg side first instead, because that's the one you'd have the information about.

Reaction mass calculations — no worries, matey...

The only way to get good at these is to practise. So make sure you can do the example, then try these:

1) Find the mass of calcium which gives 30 g of calcium oxide (CaO) when burnt in air.
2) What mass of fluorine fully reacts with potassium to make 116 g of potassium fluoride (KF)?

Answers on page 92.

Isolating the Product and Measuring Yield

Once all the boring maths has been done, chemists can crack on with making, isolating and purifying the product. Maths is never far away though — they then have to calculate the yield.

Isolating the Product and Purification Use Similar Techniques

The product needs to be isolated from the reaction mixture and purified. Here's where the real fun starts.

FILTRATION — used to separate an insoluble solid from a liquid.

1) Filtration can be used if the product is an insoluble solid that needs to be separated from a liquid reaction mixture, e.g. in the pharmaceutical industry it's used to separate out aspirin.
2) It can be used in purification as well. For example solid impurities in the reaction mixture can be separated out using filtration.

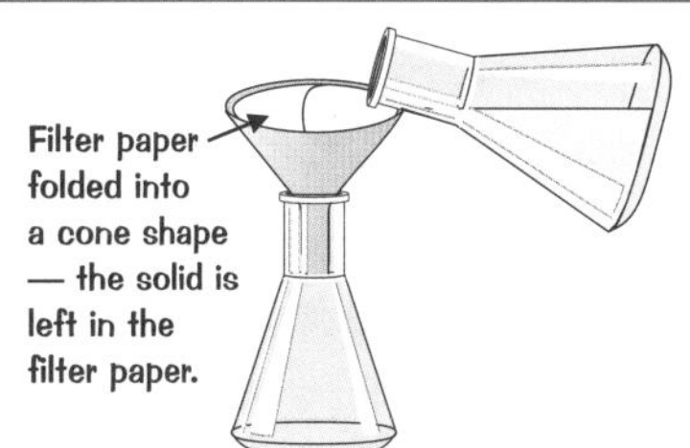
Filter paper folded into a cone shape — the solid is left in the filter paper.

EVAPORATION and CRYSTALLISATION — used to separate a soluble solid from solution.

1) Heating up the solution causes the solute to evaporate, leaving behind solid crystals of the product.
2) This is also useful for purifying the product. The crystals have a regular structure that the impurities can't fit into.
3) This process is often repeated over and over again to improve the purity. Products are dissolved and then crystallised again, which is called recrystallisation.

evaporating dish

HEAT

DRYING — used to dry the product by removing excess liquid.

1) The product can be dried in a drying oven. Some simply heat the sample, but others are more like hairdriers — they blow hot, dry air over the product.
2) Products are also dried using desiccators. These are containers that contain chemicals like silica gel that remove water from their surroundings. They help to keep the product dry.

Percentage Yield Compares Actual and Theoretical Yield

You need to understand the difference between the actual yield, the theoretical yield and the percentage yield of a product:

1) ACTUAL YIELD — this is the mass of pure, dry product. It depends on the amount of reactants you started with. The actual yield is calculated by weighing the dried product.
2) THEORETICAL YIELD — this is the maximum possible mass of pure product that could have been made using the amounts of reactants you started with. It's calculated from the balanced symbol equation and the maths you learnt on page 60.
3) PERCENTAGE YIELD — this is the actual yield of the product as a percentage of the theoretical yield.

$$\text{percentage yield} = \frac{\text{actual yield (grams)}}{\text{theoretical yield (grams)}} \times 100$$

The percentage yield will always be less than 100%. That's because some product will be lost along the way, e.g. during purification and drying.

It can all be quite dull — like watching chemicals dry...

Unfortunately, no matter how careful you are, you're not going to get a 100% yield in any reaction. You'll always get a little loss of product. In industry, people work very hard to keep waste as low as possible — so reactants that don't react first time are collected and recycled whenever possible.

Titrations

Titrations have a bad reputation — but they're not as bad as they're made out to be.

Titrations are Carried Out Using a Burette

Titrations can be used to check the purity of acidic or alkaline products (see next page).
They work using neutralisation reactions (see p.55).

1) Add a known volume of alkali to a titration flask, along with two or three drops of indicator.
2) Fill a burette with the acid. A burette is a nice fancy bit of kit:

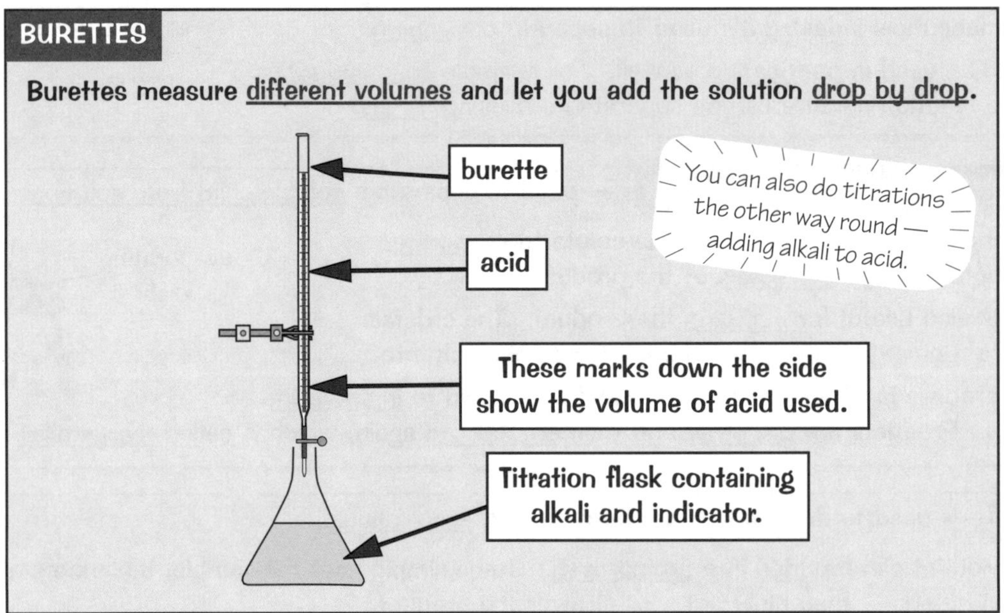

BURETTES
Burettes measure different volumes and let you add the solution drop by drop.
burette
acid
These marks down the side show the volume of acid used.
Titration flask containing alkali and indicator.
You can also do titrations the other way round — adding alkali to acid.

3) Using the burette, add the acid to the alkali a bit at a time — giving the conical flask a regular swirl. Go slow when you think the end-point (colour change) is about to be reached.
4) The indicator changes colour when all the alkali has been neutralised.
5) Record the volume of acid used to neutralise the alkali.

Solids are Weighed Out into a Titration Flask

Titrations can't be carried out with solids — only liquids. So any solid product being tested needs to be made into a solution.

1) If the product is a solid lump, it can help to first crush it into a powder.
2) Put a titration flask onto a balance.
3) Carefully weigh some of the powder into the flask. (The amount of solid used will differ from product to product.)
4) Add a solvent, e.g. water or ethanol, to dissolve the powder. (Again the solvent used and the amount will depend on what is being tested.)
5) Finally, swirl the titration flask until all of the solid has dissolved.

How do you get lean molecules? Feed them titrations...

Before the end of this module you'll be a dab hand at titrations — whether you want to be or not. They're not too tricky really — you just need to make as sure as you can that your results are accurate, which means going slowly near the end-point.

Purity

It's all very well making a product, but how do you know if it's any good or not...

Some Products Need to be Very Pure

1) The purity of a product will improve as it's being isolated. But to get a really pure product earlier stages (such as filtration, evaporation and crystallisation) will often need to be repeated.
2) Purifying and measuring the purity of a product are particularly important steps in the chemical industry. For example, in the production of...

> ...pharmaceuticals — it's really important to ensure drugs intended for human consumption are free from impurities — these could do more harm than good.
>
> ...petrochemicals — if there are impurities or contaminants in petrol products they could cause damage to a car's engine.

Titrations Can be Used to Measure the Purity of a Substance

The purity of a compound can be calculated using a titration:

Example — Determining the Purity of Aspirin

Say you start off with 0.2 g of impure aspirin dissolved in 25 cm^3 of ethanol in your conical flask. You find from your titration that it takes 9.5 cm^3 of 4 g/dm^3 NaOH to neutralise the aspirin solution.

Step 1: Work out the concentration of the aspirin solution.
You'll be given a formula, so all you have to do is stick the numbers in the right places.

$$\text{conc. of aspirin solution} = 4.5 \times \frac{\text{conc. of NaOH} \times \text{vol. of NaOH}}{\text{vol. of aspirin solution}}$$

$$= 4.5 \times \frac{4 \times (9.5 / 1000)}{25 / 1000} = \underline{6.84\ g/dm^3}$$

This number will change depending on which acid and alkali are used.

The volumes need to be in dm^3, not cm^3, so you need to divide by 1000 to convert it.

1 dm^3 = 1000 cm^3

Step 2: Work out the mass of the aspirin. You'll need to use another formula:

$$\text{mass} = \text{concentration} \times \text{volume}$$

$$= 6.84 \times 25 / 1000 = \underline{0.171\ g}$$

This is the concentration of the aspirin solution.

In the 0.2 g that you started with, 0.171 g was aspirin.

Step 3: Calculate the purity, using the formula:

$$\%\ \text{purity} = \frac{\text{calculated mass of substance}}{\text{mass of impure substance at start}} \times 100$$

$$= 0.171 \div 0.2 \times 100 = \underline{85.5\%}$$

This is the percentage purity of the aspirin.

This is the mass of aspirin that reacted in the titration.

This is the mass of impure aspirin you started with.

Concentrate when doing titrations...

This is all pretty complicated stuff, but don't worry about it too much. If you get asked to do it in the exam they'll give you all the information you need — like the formulas, concentrations and volumes.

Rates of Reaction

There's an old saying in the chemical industry — the faster you make chemicals the faster you make money. For this reason it's important to know what factors affect the rate of a chemical reaction.

Reactions Can Go at All Sorts of Different Rates

The rate of a chemical reaction is how fast the reactants are changed into products.

1) One of the slowest is the rusting of iron (it's not slow enough though — what about my little MGB).
2) Other slow reactions include chemical weathering — like acid rain damage to limestone buildings.
3) An example of a moderate speed reaction is a metal (e.g. magnesium) reacting with acid to produce a gentle stream of bubbles.
4) Burning is a fast reaction, but explosions are even faster and release a lot of gas. Explosive reactions are all over in a fraction of a second.

Controlling the Rate of Reaction is Important in Industry

In industry, it's important to control the rate of a chemical reaction for two main reasons:

1) Safety — if the reaction is too fast it could cause an explosion, which can be a bit dangerous.
2) Economic reasons — changing the conditions can be costly. For example, using very high temperatures means there'll be bigger fuel bills, so the cost of production is pushed up. But, a faster rate means that more product will be produced in less time. Companies often have to choose optimum conditions that give low production costs, but this may mean compromising on the rate of production, or the yield.

In the exam they could ask you to interpret information about rates of reaction in chemical synthesis. You might be given a load of info and asked to pick the best process. You'll have to think about what the optimum conditions would be — look for reactions that will give the best yield and rate of production for the lowest cost. You'll also need to think about environmental issues (like poisonous gases) and how dangerous the reactions are.

Typical Graphs for Rate of Reaction

The plot below shows how the speed of a particular reaction varies under different conditions. The quickest reactions have the steepest lines and become flat in the least time.

1) Graph 1 represents the original reaction.
2) Graph 2 represents the reaction taking place quicker, but with the same initial amounts. The same amount of product is produced overall — just at a quicker rate.
3) The increased rate could be due to any of these:

> a) increase in temperature
> b) increase in concentration
> c) catalyst added
> d) solid reactant crushed up into smaller bits.

Make sure you understand the graphs — you might be given data to interpret in the exam.

Amount of product evolved
④ faster, and more reactants
Flat lines show the reaction has finished.
③ much faster reaction
② faster reaction
① original reaction
Time

4) Graph 4 shows more product as well as a faster reaction. This can only happen if more reactant(s) are added at the start. Graphs 1, 2 and 3 all converge at the same level, showing that they all produce the same amount of product, although they take different times to produce it.

Get a fast, furious reaction — tickle your teacher...

First off... remember that the amount of product you get depends only on the amount of reactants you start with. So all this stuff about the rate of a reaction is only talking about how quickly your products form — not how much you get. It's an important difference — so get your head round it ASAP.

Collision Theory

Reaction rates are explained perfectly by collision theory. It's really simple.

It just says that the rate of a reaction simply depends on how often and how hard the reacting particles collide with each other. The basic idea is that particles have to collide in order to react, and they have to collide hard enough (with enough energy).

The Rate of Reaction Depends on Four Things:

1) Temperature
2) Concentration
3) Catalyst
4) Size of 'lumps' (or surface area)

LEARN THEM!

More Collisions Increases the Rate of Reaction

All four methods of increasing the rate of a reaction can be explained in terms of increasing the number of successful collisions between the reacting particles:

Increasing the TEMPERATURE Increases the Rate of Reaction

1) When the temperature is increased the particles all move faster.
2) If they're moving faster, they're going to have more collisions.
3) Also the faster they move the more energy they have, so more of the collisions will have enough energy to make the reaction happen.

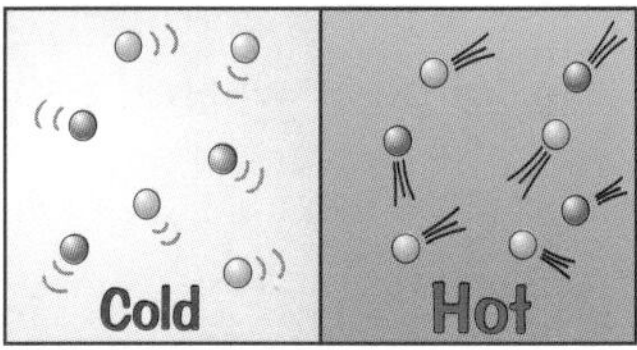

Increasing the CONCENTRATION Increases the Rate of Reaction

1) If a solution is made more concentrated it means there are more particles of reactant knocking about in the same volume of water (or other solvent).
2) This makes collisions between the reactant particles more likely.

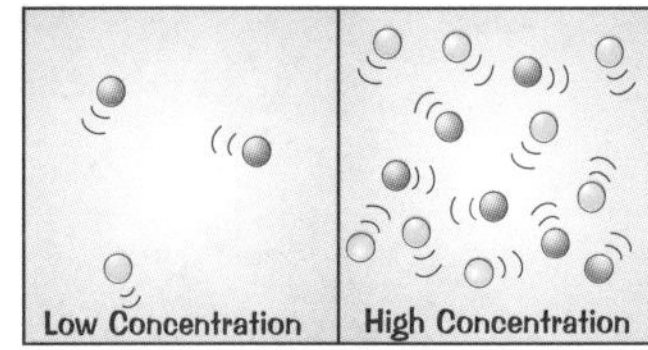

SMALLER SOLID PARTICLES (or MORE SURFACE AREA) Increases the Rate of Reaction

1) If one of the reactants is a solid then breaking it up into smaller pieces will increase its surface area.
2) This means the particles around it will have more area to work on so there'll be more collisions.
3) For example, soluble painkillers dissolve faster when they're broken into bits.

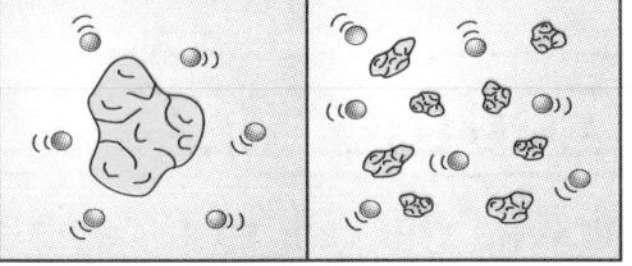

Using a CATALYST Increases the Rate of Reaction

1) A catalyst is a substance which increases the speed of a reaction, without being changed or used up in the reaction.
2) A catalyst works by giving the reacting particles a surface to stick to where they can bump into each other — increasing the number of successful collisions.

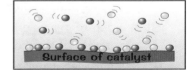

Surface of catalyst

Collision theory — it's always the other driver...

Industries that use chemical reactions to make their products have to think carefully about reaction rates. Ideally, they want to speed up the reaction to get the products quickly, but high temperatures and pressures are expensive. So they compromise — they use a slower reaction but a cheaper one.

Measuring Rates of Reaction

Three Ways to Measure the Speed of a Reaction

The speed of a reaction can be observed either by how quickly the reactants are used up or how quickly the products are formed. It's usually a lot easier to measure products forming.

The rate of reaction can be calculated using the following equation:

$$\text{Rate of Reaction} = \frac{\text{Amount of reactant used or amount of product formed}}{\text{Time}}$$

There are different ways that the speed of a reaction can be measured. Learn these three:

1) Precipitation

1) This is when the product of the reaction is a precipitate which clouds the solution.
2) Observe a mark through the solution and measure how long it takes for it to disappear.
3) The faster the mark disappears, the quicker the reaction.
4) This only works for reactions where the initial solution is see-through.
5) The result is very subjective — different people might not agree over the exact point when the mark 'disappears'.

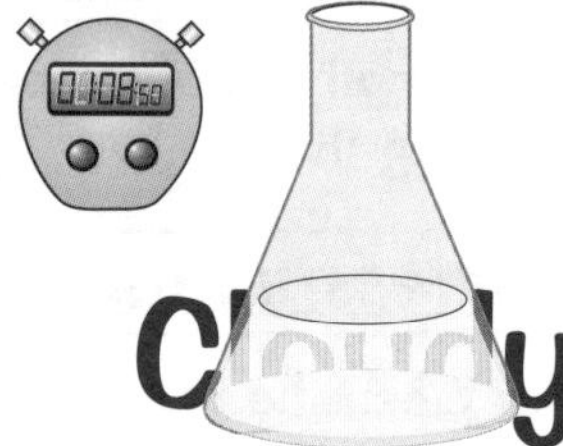

2) Change in Mass (Usually Gas Given Off)

1) Measuring the speed of a reaction that produces a gas can be carried out using a mass balance.
2) As the gas is released the mass disappearing is easily measured on the balance.
3) The quicker the reading on the balance drops, the faster the reaction.
4) Rate of reaction graphs are particularly easy to plot using the results from this method.
5) This is the most accurate of the three methods described on this page because the mass balance is very accurate. But it has the disadvantage of releasing the gas straight into the room.

3) The Volume of Gas Given Off

1) This involves the use of a gas syringe to measure the volume of gas given off.
2) The more gas given off during a given time interval, the faster the reaction.
3) A graph of gas volume against time could be plotted to give a rate of reaction graph.
4) Gas syringes usually give volumes accurate to the nearest cm^3, so they're quite accurate. You have to be quite careful though — if the reaction is too vigorous, you can easily blow the plunger out of the end of the syringe!

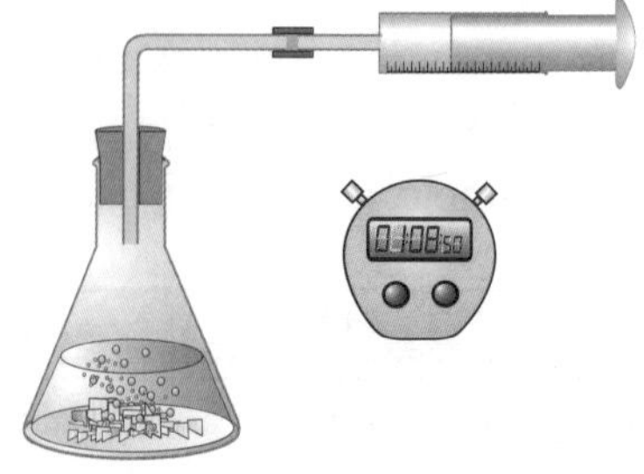

OK, have you got your stopwatch ready... *BANG!* — oh...

Each method has its pros and cons. The mass balance method is only accurate as long as the flask isn't too hot, otherwise you lose mass by evaporation as well as by the reaction. The first method isn't very accurate, but if you're not producing a gas you can't use either of the other two. Ah well.

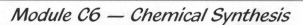

Revision Summary for Module C6

And that's it... the end of another section. Which means it's time for some more questions. There's no point in trying to duck out of these — they're the best way of testing that you've learned everything in this topic. If you can't answer any of them, look back in the book. If you can't do all this now, you won't be able to in the exam either.

1) Name two chemicals that you might come across in everyday life.
2) Give an example of one chemical that is produced on a small scale and one chemical that is produced on a large scale.
3) State whether substances with the following pH are acid, alkali or neutral:
   a) pH 2 b) pH 13 c) pH 0 d) pH 7
4) Give an example of a pure acidic compound that is a liquid.
5) Name two ways of measuring the pH of a substance.
6) What types of ions are always present in: a) acids, and b) alkalis dissolved in water?
7) Write the equation for neutralisation in terms of the ions you named in question 6.
8) Why is it dangerous to add potassium to an acid?
9) Write balanced symbol equations and name the salts formed in the following reactions:
   a) hydrochloric acid with: i) magnesium, ii) aluminium and iii) zinc,
   b) sulfuric acid with: i) magnesium, ii) aluminium and iii) zinc.
10) Name the salts formed and write a balanced equation for the reaction between:
    a) hydrochloric acid and copper oxide.
    b) hydrochloric acid and calcium hydroxide.
11) When designing a chemical process why is it important to carry out a risk assessment?
12) Why is it important to choose the right apparatus for a chemical process?
13) What does calculating the yield tell you about a reaction?
14)* Find A_r or M_r for each of these (use the periodic table):
    a) Ca b) Ag c) CO_2 d) $MgCO_3$ e) $Al(OH)_3$
    f) ZnO g) Na_2CO_3 h) sodium chloride
15)* Write down the method for calculating reacting masses.
    a) What mass of magnesium oxide is produced when 112.1 g of magnesium burns in air?
    b) What mass of sodium is needed to produce 108.2 g of sodium oxide?
16) How would you separate an insoluble product from a liquid reaction mixture?
17) Name two stages in the synthesis of a chemical where evaporation can be useful.
18) Give two methods used to dry a product.
19) What is the formula for percentage yield? How does it differ from actual yield?
20) Why is purification of a product important?
21) Describe how to carry out a titration.
22)* Calculate the purity of a 0.5 g aspirin tablet that contains 0.479 g of aspirin.
23) What four things affect the rate of a reaction?
24) Give two reasons why it is important to control the rate of a chemical reaction in industry.
25)* Magnesium metal was placed into a solution of 0.1 M hydrochloric acid. The reaction produced 50 cm^3 of hydrogen. Would you expect the same reaction with 0.2 M hydrochloric acid to be faster or slower? Explain why.
26) Describe three different ways of measuring the rate of a reaction. Give one advantage and one disadvantage of each method.

* Answers on page 92.

Alkanes

This module kicks off with some exciting organic chemistry, hooray. Organic chemistry is all about carbon compounds — it's good because everything is grouped into nice, tidy families. So, to begin, the alkanes...

Alkanes Are a Family of Hydrocarbons

1) Alkanes are made up of chains of carbon atoms surrounded by hydrogen atoms.
2) Alkanes only contain single covalent bonds (see page 45).
3) The alkane family contains molecules that look similar, but have different length chains of carbon atoms.
4) All alkanes have the formula: C_nH_{2n+2}  n is just the number of carbon atoms in the chain.

Methane, Ethane, Propane and Butane are Alkanes

The first four alkanes are methane, ethane, propane and butane.

Name	Methane	Ethane	Propane	Butane
Molecular formula	CH_4	C_2H_6	C_3H_8	C_4H_{10}
Structural formula	H–C–H with H above and below C	H–C–C–H with H above and below each C	H–C–C–C–H with H above and below each C	H–C–C–C–C–H with H above and below each C
Ball-and-stick representation				

Alkanes Burn to Give Carbon Dioxide and Water

Alkanes burn to produce carbon dioxide and water, provided there's plenty of oxygen around.

alkane + oxygen → carbon dioxide + water

You need to be able to give a balanced symbol equation for the combustion (burning) of an alkane when you're given its molecular formula. It's pretty easy — here's an example:

Make sure you end up with the same number of Cs, Hs and Os on either side of the arrow.

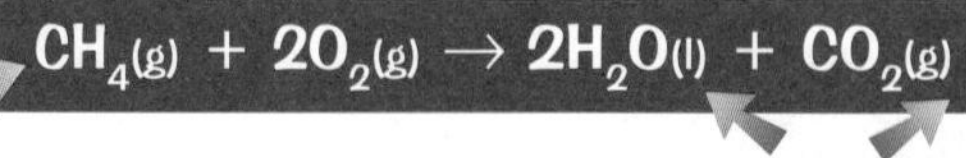

Don't forget your state symbols — s for solid, l for liquid and g for gas.

Alkanes Don't React with Most Chemicals

1) Alkanes are pretty unreactive towards most chemicals.
2) They don't react with aqueous reagents (substances dissolved in water).
3) Alkanes don't react because the C–C bonds and C–H bonds in them are unreactive.

Alkane anybody who doesn't learn this lot properly...

The clever thing about the names of alkanes is that they tell you their structure — "Meth-" means "one carbon atom", "eth-" means "two C atoms", "prop-" means "three C atoms", "but-" means "four C atoms".

Alcohols

You need to learn the structure, physical properties, chemical properties and uses of alcohols.

Alcohols Have an '-OH' Functional Group and End in '-ol'

1) The general formula for an alcohol is $C_nH_{2n+1}OH$.
2) You need to know the first two alcohols — methanol CH_3OH and ethanol C_2H_5OH.
3) The '–OH' bit is called the functional group.
4) All alcohols have similar properties because they all have the –OH functional group.

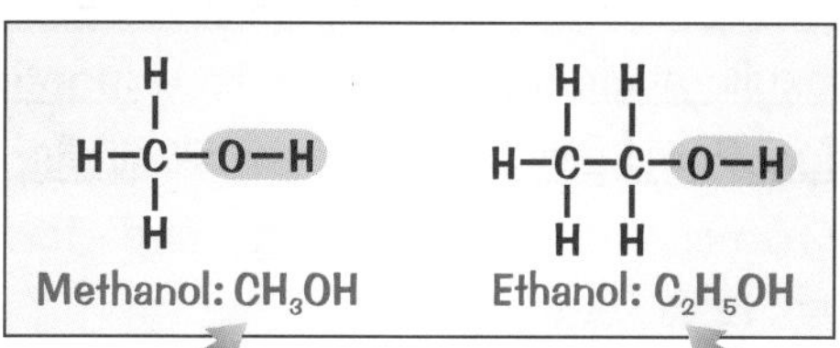

Don't write CH_4O instead of CH_3OH, or C_2H_6O instead of C_2H_5OH — it doesn't show the functional -OH group.

Alcohols, Alkanes and Water — The Similarities and Differences

You need to know how alcohols compare with alkanes and water in terms of their physical properties:

1) Ethanol is soluble in water. Alkanes are insoluble in water.
2) Ethanol and water are both good solvents — lots of things dissolve in them.
3) The boiling point of ethanol is 78 °C. This is lower than the boiling point of water (100 °C), but much higher than the boiling point of a similar size alkane (e.g. ethane has a boiling point of –103 °C).
4) Ethanol is a liquid at room temperature. It evaporates easily and gives off fumes (i.e. it's volatile). Methane and ethane are also volatile, but are gases at room temperature. Water is liquid at room temp, but not volatile.

Alcohols are Used as Solvents and Fuels and in Manufacturing

1) Alcohols, such as methanol and ethanol, can dissolve lots of compounds that water can't — e.g. hydrocarbons and oils. This makes methanol and ethanol very useful solvents in industry.
2) Methanol is also used in industry as a starting point for manufacturing other organic chemicals.
3) Ethanol is used in perfumes and aftershave lotions as it can mix with both the oils (which give the smell) and the water (that makes up the bulk).
4) 'Methylated spirit' (or 'meths') is ethanol with chemicals (e.g. methanol) added to it. It's used to clean paint brushes and as a fuel (among other things).
5) Alcohols burn in air because they contain hydrocarbon chains. Pure ethanol is clean burning so it is sometimes mixed with petrol and used as fuel for cars to conserve crude oil.

Alcohols React With Sodium

1) Sodium metal reacts gently with ethanol, to produce sodium ethoxide and hydrogen.

sodium + ethanol → sodium ethoxide + hydrogen

2) Sodium metal reacts much more vigorously with water — even melting because of the heat of the reaction.

sodium + water → sodium hydroxide + hydrogen

3) Alkanes do not react with sodium at all.

Quick tip — don't fill your car with single malt whisky...

Alcohols don't have too many chemical reactions that you need to know about for GCSE — just the two above. You do need to know the formulas, the physical properties and the uses of alcohols though. And remember — there are an awful lot more uses for alcohols than just making drinks.

Carboxylic Acids

Carboxylic acids are another happy family of organic chemicals.

Carboxylic Acids Have Functional Group -COOH

1) Carboxylic acids have '-COOH' as a functional group.
2) The functional group gives them all similar properties.
3) Their names end in '-anoic acid' (and start with the normal 'meth/eth...').

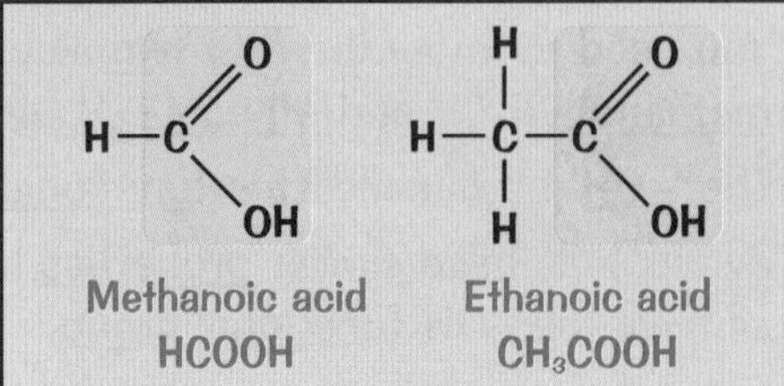

Methanoic acid HCOOH

Ethanoic acid CH_3COOH

Carboxylic Acids React Like Other Acids

1) Carboxylic acids react with alkalis, carbonates and reactive metals just like any other acid.
2) The salts formed in these reactions end in -anoate — e.g. methanoic acid forms a methanoate, ethanoic acid forms an ethanoate, etc.

Carboxylic acids react with metals to give a salt and hydrogen:

ethanoic acid + magnesium → magnesium ethanoate + hydrogen

$$2CH_3COOH_{(aq)} + Mg_{(s)} \rightarrow Mg(CH_3CO_2)_{2(aq)} + H_{2(g)}$$

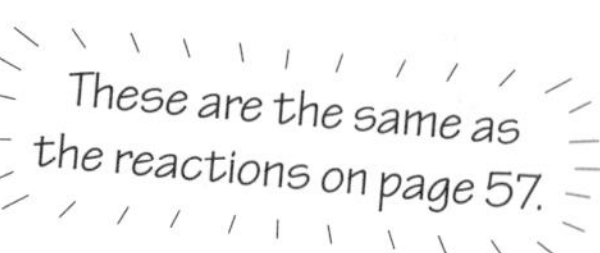

These are the same as the reactions on page 57.

Carboxylic acids react with alkalis to form a salt and water:

ethanoic acid + magnesium hydroxide → magnesium ethanoate + water

$$2CH_3COOH_{(aq)} + Mg(OH)_{2(aq)} \rightarrow Mg(CH_3CO_2)_{2(aq)} + 2H_2O_{(l)}$$

Carboxylic acids react with carbonates to give a salt, water and carbon dioxide:

ethanoic acid + magnesium carbonate → magnesium ethanoate + water + carbon dioxide

$$2CH_3COOH_{(aq)} + MgCO_{3(aq)} \rightarrow Mg(CH_3CO_2)_{2(aq)} + H_2O_{(l)} + CO_{2(g)}$$

3) Carboxylic acids are weak acids (p.75). They don't react as fast as strong acids like hydrochloric acid.

Carboxylic Acids Stink

Eugh... nasty carboxylic acids...

1) Carboxylic acids often have strong smells and tastes — they're the reason your sweaty socks stink after P.E. and why gone off (rancid) butter tastes gross.
2) If wine or beer is left open to the air, the ethanol is oxidised to ethanoic acid. This is why drinking wine after it's been open for a few days is like drinking vinegar — it is vinegar.

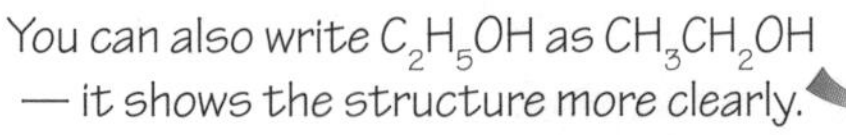

ethanol + oxygen → ethanoic acid + water

$$CH_3CH_2OH_{(aq)} + O_{2(g)} \rightarrow CH_3COOH_{(aq)} + H_2O_{(l)}$$

3) The strong smell and taste can be useful sometimes too... Vinegar is a dilute solution of ethanoic acid and tasty on your chips.

Ethanoic acid — it's not just for putting on your chips...

The reactions of carboxylic acids are easy — they react just like any acid (and have a pH less than 7). Examiners like you to be able to give some real-world uses too. Trust me... there are worse topics.

Esters

Esters are lovely things — all fruity and sweet, mmm...

Esters Have Functional Group -COO-

1) Esters are another family of organic chemicals. They all have the same functional group, –COO–.
2) They're formed from an alcohol and a carboxylic acid. It's called an esterification reaction.
3) You need to know the word equation for the reaction:

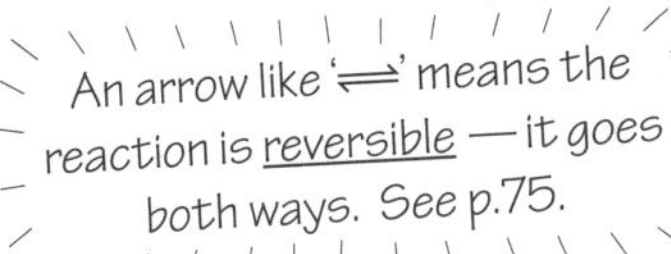

An arrow like '⇌' means the reaction is reversible — it goes both ways. See p.75.

$$\text{alcohol} + \text{carboxylic acid} \rightleftharpoons \text{ester} + \text{water}$$

Esters are Often Used in Flavourings and Perfumes

1) Many esters have pleasant smells — often quite sweet and fruity. The nice fragrances and flavours of lots of fruits come from esters.
2) They're also volatile. This makes them ideal for perfumes (the molecules evaporate easily, so they can drift to the smell receptors in your nose).
3) Esters are also used to make flavourings and aromas — e.g. there are esters that smell or taste of rum, apple, orange, banana, grape, pineapple, etc.
4) Some esters are used in ointments (they give Deep Heat its smell).
5) Other esters are used as solvents for paint, ink, glue and in nail varnish remover.
6) Esters are also used as plasticisers — they're added to plastics to make them more flexible.

Fats and Oils are Esters of Glycerol and Fatty Acids

1) Fatty acids are carboxylic acids with long chains. They often have between 16 and 20 carbon atoms.
2) Glycerol is an alcohol — notice that '-ol' at the end.
3) Fatty acids and glycerol combine to make fats and oils.
4) Most of a fat or oil molecule consists of fatty acid chains — these give them many of their properties.
5) Fatty acids can be saturated (only C–C single bonds) or unsaturated (with C=C double bonds).

Plants and Animals Make Oils and Fats to Store Energy

1) Fats have lots of energy packed into them — so they're good at storing energy.
2) When an organism has more energy than it needs it stores the extra as fat. The fat can then be used later on when the organism needs more energy.
3) The fats that plants and animals make have different properties:

Animal fats have mainly saturated hydrocarbon chains. They contain very few C=C bonds. They are normally solid at room temperature.

Vegetable oils have mainly unsaturated hydrocarbon chains. They contain lots of C=C bonds. They are normally liquid at room temperature.

What's a chemist's favourite chocolate — ester eggs...

If you want to impress your friends learn the full name of glycerol — propane-1,2,3-triol — then just casually drop it into conversation, as in "On Saturday I... err..." well, good luck with that one.

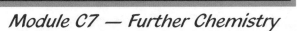

Making an Ester

Take an alcohol from p.69, mix it with an acid from p.70, and what have you got... an ester, that's what.

How to Make an Ester — Reflux, Distil, Purify, Dry

Making esters is a little more complicated than just mixing an alcohol and a carboxylic acid together. The reaction is reversible so some of the ester formed will react with the water produced and re-form the carboxylic acid and alcohol. To get a pure ester you need a multi-step reaction and purification procedure.

1) Refluxing — The Reaction

To make ethyl ethanoate you need to react ethanol with ethanoic acid, using a catalyst to speed things up (concentrated sulfuric acid is a good choice).

Heating the mixture also speeds up the reaction — but you can't just stick a Bunsen under it as the ethanol will evaporate or catch fire before it can react.

Instead, the mixture's gently heated in a flask fitted with a condenser — this catches the vapours and recycles them back into the flask, giving them time to react. This handy method is called refluxing.

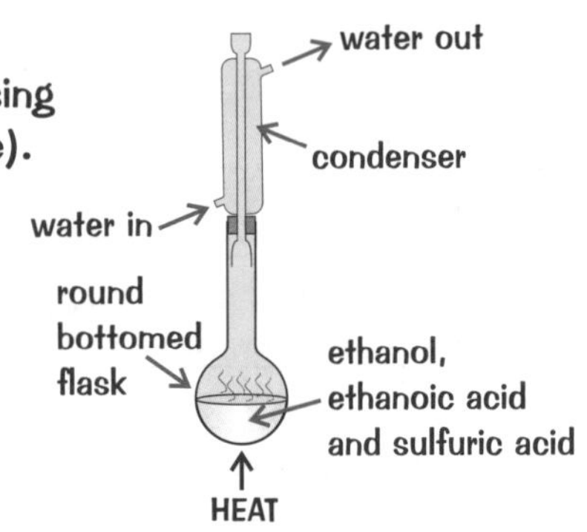

2) Distillation

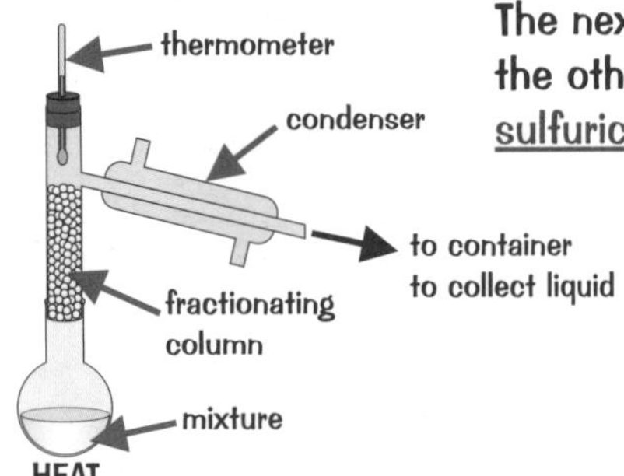

The next step is distillation. This separates your lovely ester from all the other stuff left in the flask (unreacted alcohol and carboxylic acid, sulfuric acid and water).

The mixture's heated below a fractionating column. As it starts to boil, the vapour goes up the fractionating column.

When the temperature at the top of the column reaches the boiling point of ethyl ethanoate, the liquid that flows out of the condenser is collected. This liquid is impure ethyl ethanoate.

3) Purification

The liquid collected (the distillate) is poured into a tap funnel and then treated to remove its impurities, as follows:

The mixture is shaken with sodium carbonate solution to remove acidic impurities. Ethyl ethanoate doesn't mix with water, so the mixture separates into two layers, and the lower layer can be tapped off (removed).

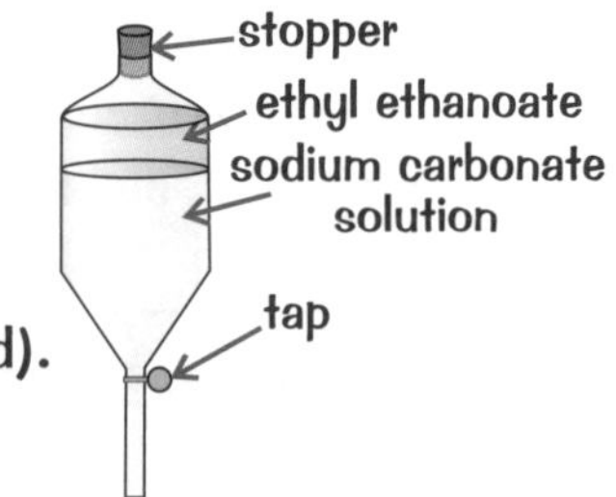

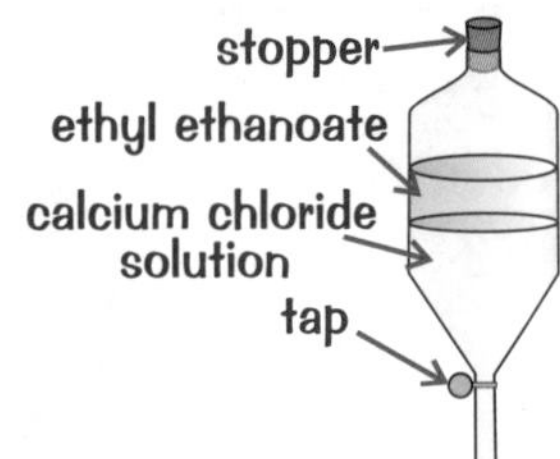

The remaining upper layer is then shaken with concentrated calcium chloride solution to remove any ethanol. Again, the lower layer can be tapped off and removed.

4) Drying

Any remaining water in the ethyl ethanoate can be removed by shaking it with lumps of anhydrous calcium chloride, which absorb the water — this is called drying. Finally, the pure ethyl ethanoate can be separated from the solid calcium chloride by filtration.

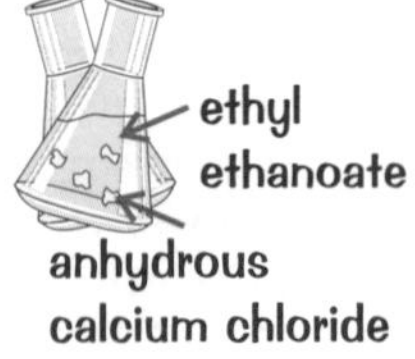

Esterification — it's 'esterical stuff...

Sorry about that, it's a little bit tricky really. The secret is to get your head around each step of the process before you try and put it all together. Read each step, cover, scribble... you know the drill.

Energy Transfer in Reactions

Whenever chemical reactions occur, there are changes in energy. This is kind of interesting if you think of the number of chemical reactions that are involved in everyday life.

Reactions are Exothermic or Endothermic

An EXOTHERMIC reaction is one which gives out energy to the surroundings, usually in the form of heat and usually shown by a rise in temperature.

E.g. fuels burning or neutralisation reactions.

An ENDOTHERMIC reaction is one which takes in energy from the surroundings, usually in the form of heat and usually shown by a fall in temperature.

E.g. photosynthesis.

Energy Level Diagrams Show if it's Exo- or Endothermic

In exothermic reactions ΔH is –ve

1) This shows an exothermic reaction — the products are at a lower energy than the reactants.
2) The difference in height represents the energy given out in the reaction. ΔH is –ve here.

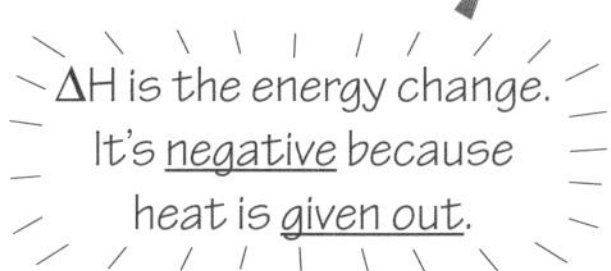

ΔH is the energy change.
It's negative because
heat is given out.

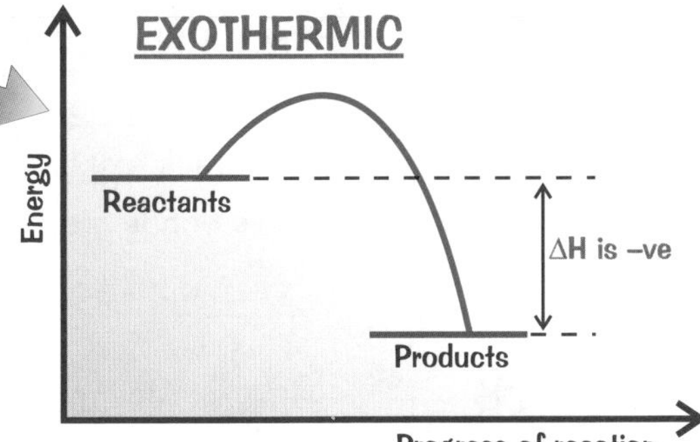

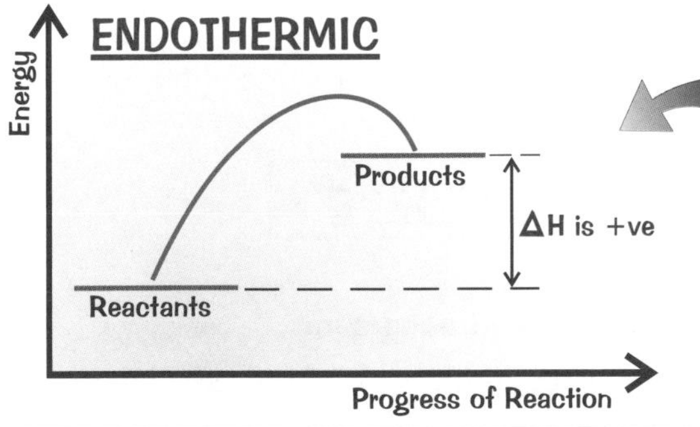

In endothermic reactions ΔH is +ve

1) This shows an endothermic reaction because the products are at a higher energy than the reactants, so ΔH is +ve.
2) The difference in height represents the energy taken in during the reaction.

Activation Energy is the Energy Needed to Start a Reaction

1) The activation energy is the minimum amount of energy needed for a reaction to happen.
2) It's a bit like having to climb up one side of a hill before you can ski/snowboard/sledge/fall down the other side.
3) If the energy input is less than the activation energy there won't be enough energy to start the reaction — so nothing will happen.

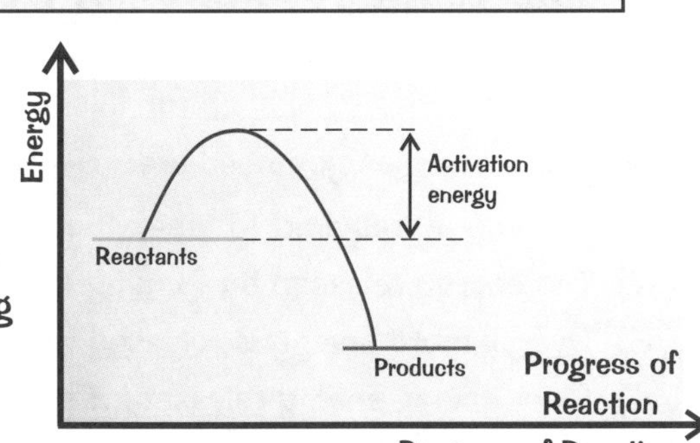

Progress of Reaction

Right, so burning gives out heat — really...

This whole energy transfer thing is a fairly simple idea — don't be put off by the long words. Remember, "exo-" = exit, "-thermic" = heat, so an exothermic reaction is one that gives out heat. And "endo-" = erm... the other one. Okay, so there's no easy way to remember that one. Tough.

Catalysts and Bond Energies

Energy transfer in chemical reactions is all to do with making and breaking bonds.

Catalysts Lower the Activation Energy

1) A catalyst is a substance which changes the speed of a reaction, without being changed or used up in the reaction.
2) Catalysts lower the activation energy needed for reactions to happen by providing alternative routes.
3) The effect of a catalyst is shown by the lower curve on the diagram.
4) The overall energy change for the reaction, ΔH, remains the same though.

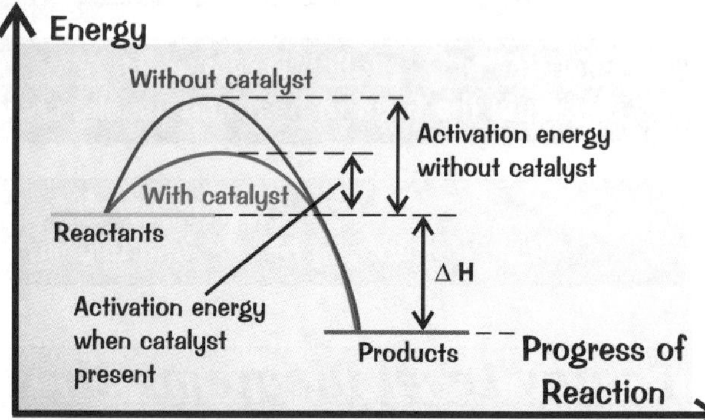

Energy Must Always be Supplied to Break Bonds

1) During a chemical reaction, old bonds are broken and new bonds are formed.
2) Energy must be supplied to break existing bonds — so bond breaking is an endothermic process.
3) Energy is released when new bonds are formed — so bond formation is an exothermic process.

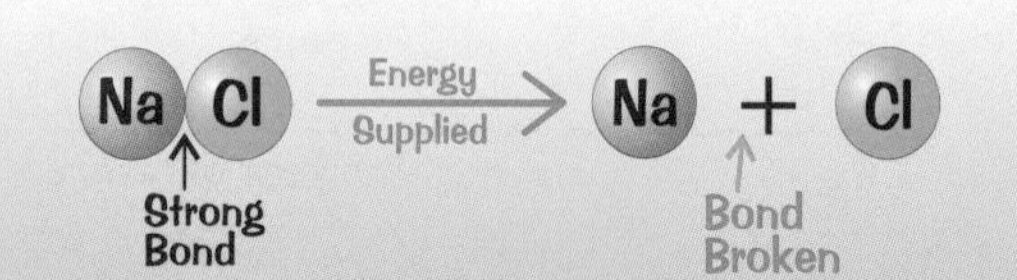

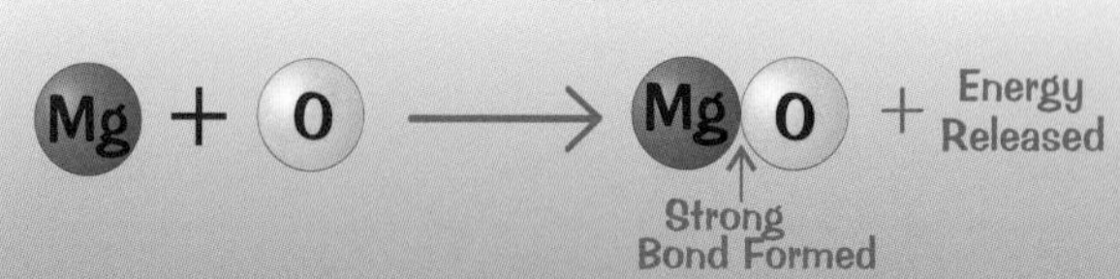

In exothermic reactions the energy released by forming bonds is greater than the energy used to break them.
In endothermic reactions the energy used to break bonds is greater than the energy released by forming them.

Bond Energy Calculations — Need to be Practised

1) Every chemical bond has a particular bond energy associated with it. This bond energy varies slightly depending on the compound the bond occurs in.
2) You can use these known bond energies to calculate the overall energy change for a reaction. You need to practise a few of these, but the basic idea is really very simple...

Example: The Formation of HCl

Using known bond energies you can calculate the energy change for this reaction:

$$\text{H—H} + \text{Cl—Cl} \longrightarrow \begin{matrix}\text{H—Cl}\\ \text{H—Cl}\end{matrix}$$

The bond energies you need are: H—H: +436 kJ; Cl—Cl: +242 kJ; H—Cl: +431 kJ.

1) The energy required to break the original bonds is 436 + 242 = +678 kJ
2) The energy released by forming the new bonds is 2 × 431 = +862 kJ
3) Overall more energy is released than is used to form the products: 862 – 678 = 184 kJ released.
4) Since this is energy released, if you wanted to show ΔH you'd need to put a negative sign in front of it to indicate that it's an exothermic reaction, like this: ΔH = –184 kJ

Energy transfer — make sure you take it all in...

Bond energies don't tell you the amount of energy associated with a single bond. They tell you the amount of energy associated with 6×10^{23} bonds. This might seem a little bit odd, but it makes sense to the clever chemistry types. In the exam you'll be given all the bond energies you'll need to use.

Reversible Reactions

A reversible reaction is one where the products of the reaction can react with each other and convert back to the original reactants. In other words, it can go both ways.

A reversible reaction is one where the products of the reaction can themselves react to produce the original reactants.

$$A + B \rightleftharpoons C + D$$

This is the symbol for a reversible reaction.

Reversible Reactions Will Reach Dynamic Equilibrium

1) If a reversible reaction takes place in a closed system then a state of equilibrium will always be reached. (A 'closed system' just means that none of the reactants or products can escape.)
2) Equilibrium means that the relative (%) quantities of reactants and products will reach a certain balance and stay there.
3) It is in fact a dynamic equilibrium, which means that the reactions are still taking place in both directions, but the overall effect is nil because the forward and reverse reactions cancel each other out. The reactions are taking place at exactly the same rate in both directions.

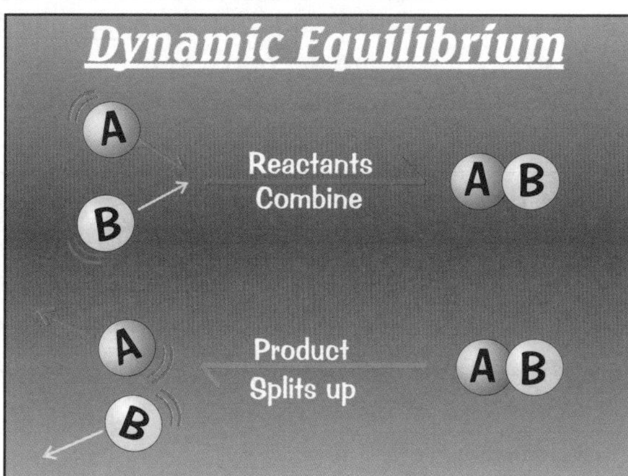

Ionisation of Weak Acids is a Reversible Reaction

When acids are dissolved in water they ionise — they release hydrogen ions, H^+. This is what makes them acidic (see p.55). For example,

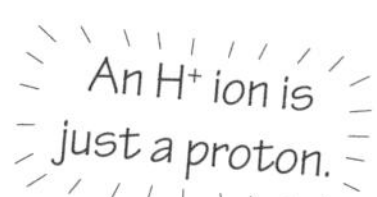

$$HCl_{(aq)} \rightarrow H^+_{(aq)} + Cl^-_{(aq)}$$

$$H_2SO_{4(aq)} \rightarrow 2H^+_{(aq)} + SO_4^{2-}{}_{(aq)}$$

Acids either ionise completely or reach a dynamic equilibrium — it all depends on the type of acid.

1) Strong acids (e.g. hydrochloric acid) ionise almost completely in water. This means almost every hydrogen is released — so there are loads of H^+ ions.
2) Weak acids (e.g. carboxylic acids) ionise only very slightly. Only some of the hydrogens in the compound are released — so only small numbers of H^+ ions are formed.

   For example,

   Strong acid: $HCl_{(aq)} \longrightarrow H^+_{(aq)} + Cl^-_{(aq)}$

   Weak acid: $CH_3COOH_{(aq)} \rightleftharpoons H^+_{(aq)} + CH_3COO^-_{(aq)}$

   Use a 'reversible reaction' arrow for a weak acid.

3) The ionisation of a weak acid is a reversible reaction. Since only a few H^+ ions are released, the equilibrium is well to the left (i.e. there's a lot more CH_3COOH molecules in the solution than there are H^+ and CH_3COO^- ions).

Keep going back and forth over reversible reactions...

...and they'll soon make sense. Acids are acidic because of H^+ ions. Strong acids are strong because they let go of all their H^+ ions at the drop of a hat... well, at the drop of a drop of water. This is tricky, it's true, but if you can get your head round this, then you can probably cope with just about anything.

Analytical Procedures

The next few pages are all about finding out exactly what is contained in a mystery substance.

Qualitative Analysis Tells You What a Sample Contains

1) Qualitative analysis tells you which substances are present in a sample.
2) It doesn't tell you how much of each substance there is — that's where quantitative analysis comes in.

Quantitative Analysis Tells You How Much it Contains

1) Quantitative analysis tells you how much of a substance is present in a sample.
2) It can be used to work out the molecular formula of the sample. E.g. if you had a sample containing carbon and hydrogen you'd know it was a hydrocarbon, but without quantitative analysis, you won't know if it's methane, butane or even 3,4-dimethylheptane...

Chemical Analysis is Carried Out On Samples

1) You usually analyse just a sample of the material under test. There's quite a few reasons for this.
2) It might be very difficult to test all of the material if you've got an awful lot of it — or you might want to just test a small bit so that you can use the rest for something else.

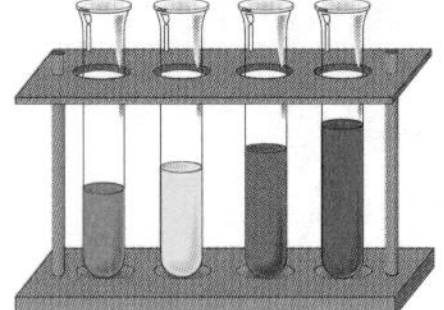

3) Taking a sample also means that if something goes wrong with the test, you can go back for another sample and try again.
4) A sample must represent the bulk of the material being tested — it wouldn't tell you anything very useful if it didn't.

Samples are Analysed in Solution

Samples are usually tested in solution. A solution is made by dissolving the sample in a solvent. There are two types of solution — aqueous and non-aqueous. Which type of solution you use depends on the type of substance you're testing.

An aqueous solution means the solvent is water. They're shown by the state symbol (aq).

A non-aqueous solution means the solvent is anything other than water — e.g. ethanol.

Standard Procedures Mean Everyone Does Things the Same Way

Whether testing chemicals or measuring giraffes, scientists follow 'standard procedures' — clear instructions describing exactly how to carry out these practical tasks.

1) Standard procedures are agreed methods of working — they are chosen because they're the safest, most effective and most accurate methods to use.
2) Standard procedures can be agreed within a company, nationally, or internationally.
3) They're useful because wherever and whenever a test is done, the result should always be the same — it should give reliable results every time.
4) There are standard procedures for the collection and storage of a sample, as well as how it should be analysed.

Analysis — don't they do that on Match of the Day?...

If you're trying to detect a certain substance in a sample, then there has to be a reasonable amount of the stuff you're looking for. You'd stand no chance of finding one molecule in a great big cake.

Analysis — Chromatography

Chromatography is one analysis method that you need to know inside out and upside down... read on.

Chromatography uses Two Phases

Chromatography is an analysis method that's used to separate the substances in a mixture. You can then use it to identify the individual substances.

There are lots of different types of chromatography — but they all have two 'phases':

- A mobile phase — where the molecules can move. This is always a liquid or a gas.
- A stationary phase — where the molecules can't move. This can be a solid or a really thick liquid.

1) The components in the mixture separate out as the mobile phase moves across the stationary phase.
2) How quickly a chemical moves depends on how it "distributes" itself between the two phases — this is why different chemicals separate out and end up at different points (see below).
3) The molecules of each chemical constantly move between the mobile and the stationary phases.
4) They are said to reach a "dynamic equilibrium" — at equilibrium the amount leaving the stationary phase for the mobile phase is the same as the amount leaving the mobile phase for the stationary phase. But be careful, this doesn't (necessarily) mean there is the same amount of chemical in each phase.

In Paper Chromatography the Stationary Phase is Paper

1) In paper chromatography, a spot of the substance being tested is put onto a baseline on the paper.
2) The bottom of the paper is placed in a beaker containing a solvent, such as ethanol or water. The solvent is the mobile phase.
3) The stationary phase is the chromatography paper (often filter paper).

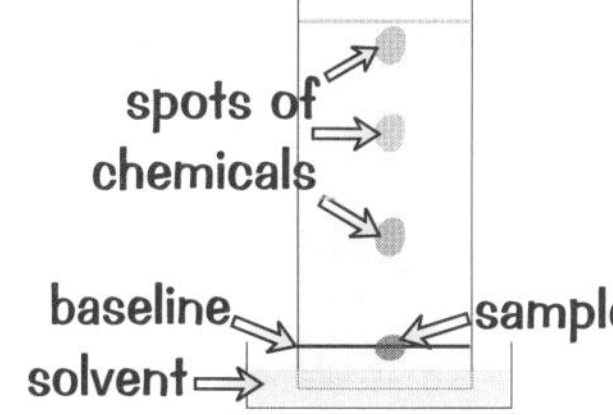

Here's what happens:

1) The solvent moves up the paper.
2) The chemicals in the sample dissolve in the solvent and move between it and the paper. This sets up an equilibrium between the solvent and the paper.
3) When they're in the mobile phase the chemicals move up the paper with the solvent.
4) Before the solvent reaches the top of the paper, the paper is removed from the beaker.
5) The different chemicals in the sample form separate spots on the paper. The chemicals that spend more time in the mobile phase than the stationary phase form spots further up the paper.

The amount of time the molecules spend in each phase depends on two things:

- how soluble they are in the solvent,
- how attracted they are to the paper.

So molecules with a higher solubility in the solvent, and which are less attracted to the paper, will spend more time in the mobile phase — and they'll be carried further up the paper.

Thin-Layer Chromatography has a Different Stationary Phase

1) Thin-layer chromatography (TLC) is very similar to paper chromatography, but the stationary phase is a thin layer of solid — e.g. silica gel spread onto a glass plate.
2) The mobile phase is a solvent such as ethanol (just like in paper chromatography).

Learning about this — it's just a phase you go through...

The tricky thing about understanding how chromatography works is that you can't see the chemicals moving between the two phases — you'll just have to believe that it does happen.

Analysis — Chromatography

You can Calculate the R_f Value for Each Chemical

1) The result of chromatography analysis is called a chromatogram.
2) Some of the spots on the chromatogram might be colourless. If they are, you need to use a locating agent to show where they are, e.g. you might have to spray the chromatogram with a reagent.
3) You need to know how to work out the R_f values for spots (solutes) on a chromatogram.
   An R_f value is the ratio between the distance travelled by the dissolved substance and the distance travelled by the solvent. You can find them using the formula:

$$R_f = \frac{\text{distance travelled by substance}}{\text{distance travelled by solvent}}$$

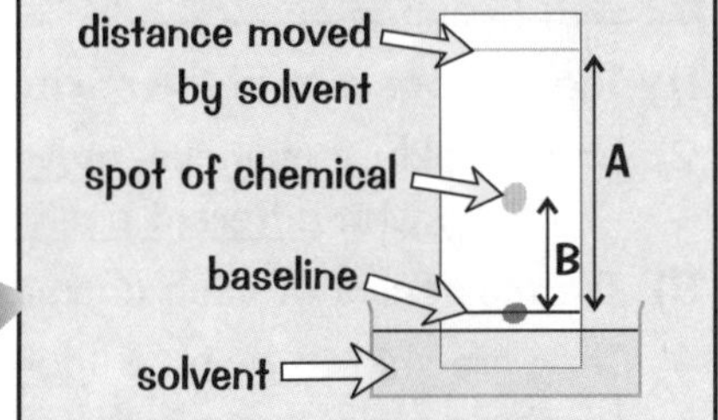

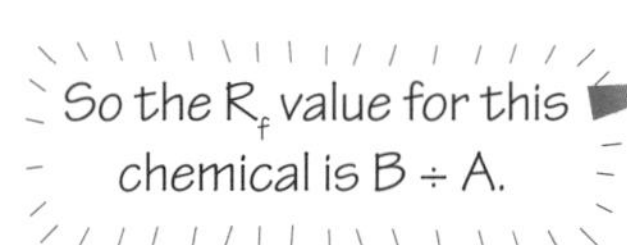

So the R_f value for this chemical is B ÷ A.

4) Chromatography is often carried out to see if a certain substance is present in a mixture. You run a pure, known sample of the substance alongside the unknown mixture. If the R_f values match, the substances may be the same (although it doesn't definitely prove they are the same).
5) Chemists use substances called standard reference materials (SRMs) to check the identities of substances. These have carefully controlled concentrations and purities.

Gas Chromatography is a Bit More High-Tech

Gas chromatography (GC) is used to analyse unknown substances too. If they're not already gases, then they have to be vaporised.

- The mobile phase is an unreactive gas such as nitrogen.
- The stationary phase is a viscous (thick) liquid, such as an oil.

mobile phase enters here
sample injected
tube coated with stationary phase
detector
temperature controlled oven

The process is quite different from paper chromatography and TLC:

1) The unknown mixture is injected into a long tube coated on the inside with the stationary phase.
2) The mixture moves along the tube with the mobile phase until it comes out the other end. Like in the other chromatography methods, the substances are distributed between the phases.
3) The time it takes a chemical to travel through the tube is called the retention time.
4) The retention time is different for each chemical — it's what's used to identify it.

The chromatogram from gas chromatography is a graph. Each peak on the graph represents a different chemical.

- The distance along the x-axis is the retention time — which can be looked up to find out what the chemical is.
- The area under the peak shows you how much of that chemical was in the sample.

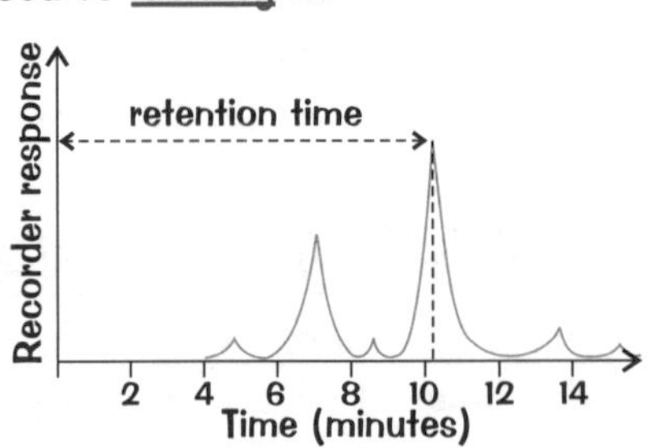

Comb-atography — identifies mysterious things in your hair...

Chromatography works by showing how mystery chemicals get distributed between mobile and stationary phases — that's what the R_f value represents. All chemicals get distributed differently, so that's how you can tell which is which. It's great — all you need is some paper and a bit of solvent.

Analysis — Solution Concentrations

A rather dull and boring page I'm afraid. But at least there are some calculations on it. Yay.

Concentration = Mass ÷ Volume

The concentration of a solution is measured in grams per dm^3 (i.e. grams per litre — one dm^3 is a litre). So 1 gram of stuff in 1 dm^3 of solution has a concentration of 1 gram per dm^3 (or 1 g/dm^3).

There's a nice formula to work out the concentration of a solution:

concentration = mass of solute ÷ volume of solution

Concentration (in g/dm^3) — Mass (in g) — Volume (in dm^3)

m
c × V

Make sure you know how to use it — you might need to rearrange it:

Example 1: 25 g of copper sulfate is dissolved in 500 cm^3 of water. What's the concentration in g/dm^3?

Answer: Make sure the amounts are in the right units — mass in g and volume in dm^3.
Substitute the values into the formula: concentration = 25 g ÷ 0.5 dm^3 = 50 g/dm^3

Convert the volume to dm^3 by dividing by 1000.

Example 2: What mass of sodium chloride is in 300 cm^3 of solution with a concentration of 12 g/dm^3?

Answer: Rearrange the formula using the triangle: mass = concentration × volume.
Substitute the values into the formula: mass = 12 g/dm^3 × 0.3 dm^3 = 3.6 g

A Standard Solution Has a Known Concentration

A standard solution is any solution that you know the concentration of.
Making a standard solution needs careful measuring and a hint of maths:

Example: Make 250 cm^3 of a 314 g/dm^3 solution of sodium chloride.

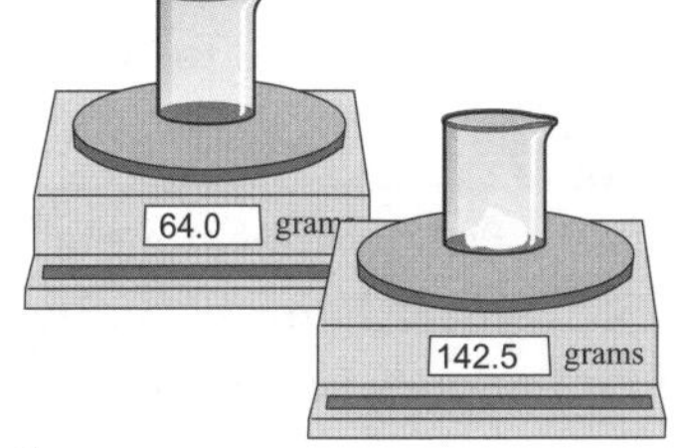

1) First work out how many grams of solute you need by using the formula: mass = concentration × volume
$= 314\ g/dm^3 \times 0.25\ dm^3 = 78.5\ g$

Remember — convert cm^3 to dm^3

2) Carefully weigh out this mass of solute — first weigh the beaker, note the weight, then add the correct mass.

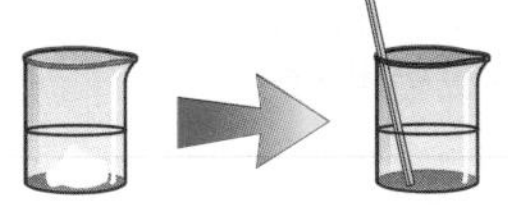

3) Add a small amount of distilled water to the beaker and stir until all the solute has dissolved.
4) Tip the solution into a volumetric flask — make sure it's the right size for the volume you're making. Use a funnel to make sure it all goes in.

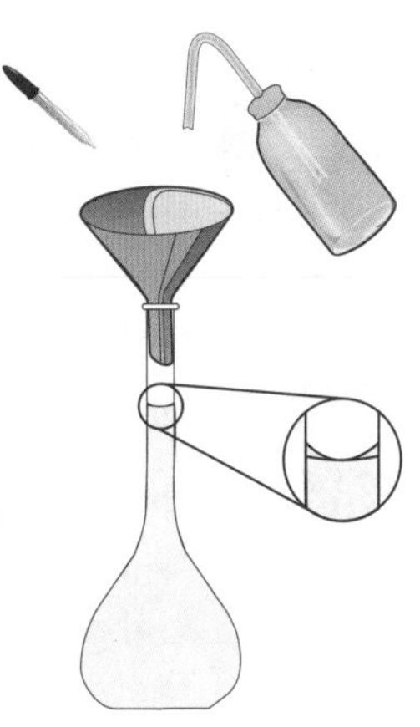

5) Rinse the beaker and stirring rod with distilled water and add that to the flask too. This makes sure there's no solute clinging to the beaker or rod.
6) Now top the flask up to the correct volume (250 cm^3) with more distilled water. Make sure the bottom of the meniscus reaches the line — when you get close to the line use a pipette to add water drop by drop. If you go over the line you'll have to start all over again.
7) Stopper the bottle and turn it upside down a few times to make sure it's all mixed.
8) Check the meniscus again and add a drop or two of water if you need to.

Wondering what's on telly? — no, don't lose concentration...

A high concentration is like a rugby team in a mini. Or everyone in Britain living on the Isle of Wight.
A low concentration is like a guy stranded on a desert island, or a small fish in a big lake. Poetic, no?

Analysis — Titration

The basics of titrations are on page 62. Here's a bit more that you need to know about them though.

You Need Several Consistent Readings

In a titration, you record the volume of acid (or alkali) added from a burette to neutralise the alkali (or acid). Sometimes you get a weird result (called an anomalous result) — this might be caused by faulty equipment, or human error (perhaps the scale was read incorrectly) — so it's best to repeat the titration a few times. If your values are all very similar you can be confident your results are reliable. If they're more spread out, you can't be so certain of what the 'true' result should be.

- The first titration should be a rough titration to get an approximate idea of the end-point.
- You then need to repeat the whole thing carefully a few times, making sure you get (pretty much) the same answer each time (within about 0.2 cm^3).
- A mean (average) value can then be calculated from the repeats — but ignore any anomalous results.

EXAMPLE A titration was repeated four times. This table shows the results:

The second result is very different from the others — it's anomalous. But the other three results are close together. You can be pretty confident that the actual result lies close to 22.3 and 22.4 cm^3.

Titration	1	2	3	4
Volume added (cm^3)	22.3	30.0	22.4	22.3

Work out the mean of the results that are close together and use it for any later calculations or graphs.

mean volume = (22.3 + 22.4 + 22.3) ÷ 3 = 22.33 cm^3 (to 2 d.p.)

Interpreting the Results of a Titration

Titrations are also used to measure purity, see p.63.

You can use titrations to work out the identity of unknown element in a compound...

Example

25 cm^3 of an unknown metal hydroxide solution (MOH) with a concentration of 192 g/dm^3 has been titrated with 40 cm^3 of hydrochloric acid solution with a concentration of 182.5 g/dm^3.
Here's the equation for the reaction: $HCl + MOH \rightarrow MCl + H_2O$
Determine what the metal hydroxide is. (Relative atomic masses: H = 1, O = 16, Cl = 35.5)

You want to find out the relative formula mass for MOH, so you can figure out what M stands for.

Step 1: First find out the mass of acid and the mass of alkali that react.

There are 192 g of MOH in each dm^3. So in 0.025 dm^3 (25 cm^3 ÷ 1000) there is 0.025 × 192 = 4.8 g of MOH.

There's 182.5 g of HCl in each dm^3. So in 0.04 dm^3, there is 0.04 × 182.5 = 7.3 g of HCl.

Step 2: Find the relative formula mass of the known solute. M_r of HCl = 1 + 35.5 = 36.5.

Step 3: Find the relative formula mass of the unknown solute using the balanced equation.

From the equation, you know that 1 molecule of HCl reacts with 1 molecule of MOH.
And you know that 7.3 g of HCl reacts with 4.8 g of MOH.
The relative formula masses tell you how the mass of the molecules of HCl and MOH compare to each other. So you can use this formula to find the relative formula mass (M_r) of MOH:

mass of HCl ÷ M_r of HCl = mass of MOH ÷ M_r of MOH

7.3 ÷ 36.5 = 4.8 ÷ M_r of MOH ⇒ M_r of MOH = 24

Step 4: Identify the metal hydroxide. M_r of MOH = (A_r of M) + 16 + 1 = 24, so A_r of M = 7.
Lithium has a relative atomic mass of 7, so the metal hydroxide, MOH, is lithium hydroxide, LiOH.

More numbers? — might as well be doing maths... urrgh...

In titration calculations, look at the information you're given, and see what you can work out with it.

The Chemical Industry

Absolutely loads of some types of chemicals are used — such as fertilisers. Well they don't just grow on trees. No, they have to be made. And made they are — on a massive scale.

Some Chemicals are Produced on a Large Scale...

There are certain chemicals that industries need thousands and thousands of tonnes of every year — ammonia, sulfuric acid, sodium hydroxide and phosphoric acid are four examples you should know. Chemicals like these that are produced on a large scale are called bulk chemicals.

...And Some are Produced on a Small Scale

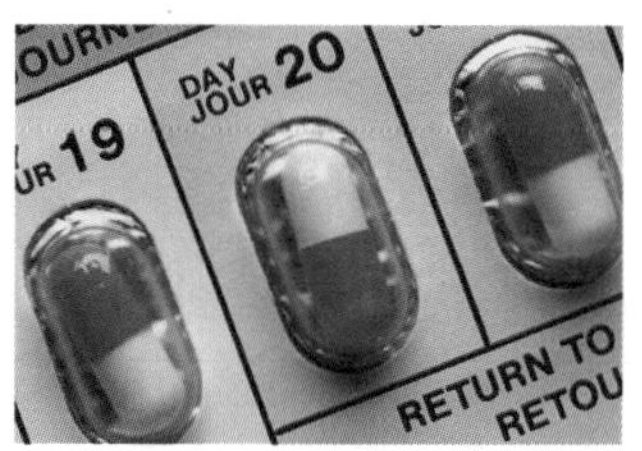

Some chemicals aren't needed in such large amounts — but that doesn't mean they're any less important.

Chemicals produced on a smaller scale are called fine chemicals. Some examples to learn are drugs, food additives and fragrances.

New Chemical Products Need Lots of Research

Before new chemical products are made, a huge amount of research and development work goes on. This can take years, and be really expensive, but it's worth it in the end if the company makes lots of money out of the new product.

For example, to make a new product efficiently a new catalyst might have to be found. This is likely to involve:

1) Testing potential catalysts using a process of trial and error.
2) Making computer models of the reaction to try to work out which substance might work as a catalyst.
3) Designing or refining the manufacture of the catalyst to make sure that the new product can be mass-produced safely, efficiently, and cost effectively.
4) Investigating the risks to the environment of using the new catalyst and trying to minimise them.
5) Monitoring the quality of the product to make sure that it is not affected by the catalyst.

These jobs, and lots of other types of work, are done by people in the chemical industry. You don't need to know about all of them off-by-heart, but you do need to be able to interpret information about them.

Government Regulations Protect People and the Environment

Governments place strict controls on everything to do with chemical processes. This is done to protect workers, the general public and the environment. For example, there are regulations about...

1) Using chemicals — e.g. sulfuric acid is sprayed on potato fields to destroy the leaves and stalks of the potato plants and make harvesting easier. Government regulations restrict how much acid can be used and require signs to be displayed to warn the public.
2) Storage — many dangerous chemicals have to be stored in locked storerooms. Noxious (poisonous) chemicals must be stored in either sealed containers or well-ventilated store cupboards.
3) Transport — e.g. lorries transporting chemicals must display hazard symbols and identification numbers to help the emergency services deal safely with any accidents and spills.

Fine chemicals — by appointment to Her Majesty, The Queen...

Producing bulk chemicals is like painting a house — it's huge, so you slap that paint on with a great big brush. Producing fine chemicals is like painting a picture — much smaller and more fiddly. Got it?

Producing Chemicals

Producing chemicals is a complicated business. Luckily most processes involve the same stages.

There Are Several Stages Involved in Producing Chemicals

The process of producing a useful chemical from the raw materials can be split into five stages:

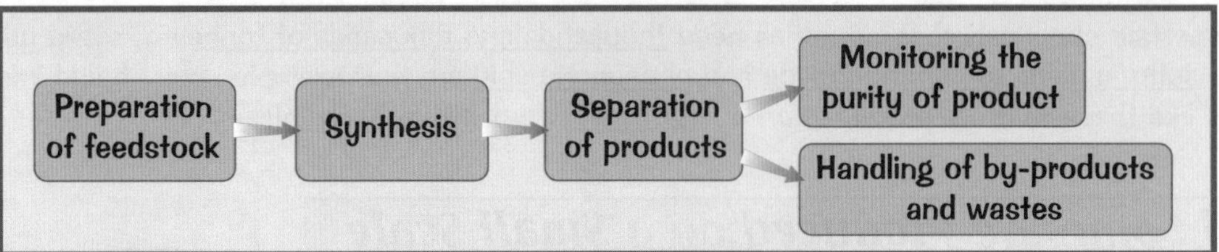

1) Raw Materials Are Converted Into Feedstocks

1) Raw materials are the naturally occurring substances which are needed, e.g. crude oil, natural gas.
2) Feedstocks are the actual reactants needed for the process, e.g. hydrogen, ethanol.
3) The raw materials usually have to be purified or changed in some way to make the feedstock.

2) Synthesis

The feedstocks (reactants) are converted by the magic of chemistry into products. The conditions have to be carefully controlled to make sure the reaction happens, and at a sensible rate.

3) The Products are Separated

1) Chemical reactions usually produce the substance you want and some other chemicals known as by-products. The by-products might be useful, or they might be waste.
2) You might also have some left-over reactants.
3) Everything has to be separated out so it can be dealt with in different ways.

There are a few other stages involved, see p.58.

4) The Purity of the Product is Monitored

1) Even after the best efforts are made to separate the product from everything else, it sometimes still has other things mixed in with it — it's not completely pure.
2) The purity of the product has to be monitored to make sure it's between certain levels.
3) Different industries need different levels of purity depending on what the product is used for. If a slightly impure product will do the job it's meant for, there's no point wasting money on purification.

5) By-products and Waste are Dealt With

1) Where possible, by-products are sold or used in another reaction.
2) If the reaction is exothermic, there may be waste heat. Heat exchangers can use excess heat to produce steam or hot water for other reactions — saving energy and money.
3) Waste products have to be carefully disposed of so they don't harm people or the environment — there are legal requirements about this.

Feedsock — when you spill your dinner over your feet...

As you might have noticed, producing chemicals isn't the most exciting of topics — but it is important. The chemical industry in Britain alone is worth billions of pounds... which makes it a lot more interesting.

Producing Chemicals

It'd be great if all industrial reactions were sustainable — humans could go on and on making whatever they wanted forever and ever. Life isn't like that though — so it's important to think about sustainability.

There Are Eight Key Questions About Sustainability

Sustainable processes are ones that meet people's needs today without affecting the ability of future generations to meet their own needs. Lots of factors affect whether a chemical process is sustainable.

1 WILL THE RAW MATERIALS RUN OUT?

It's great if your feedstock is renewable — you can keep on using as much as you like. The trouble is, if it's not renewable it's going to run out. And this could mean big problems for future generations.

2 HOW GOOD IS THE ATOM ECONOMY?

The atom economy of a reaction tells you how much of the mass of the reactants ends up as useful products. Pretty obviously, if you're making lots of waste, that's a problem — it all has to go somewhere. Reactions with low atom economy use up resources very quickly too.

3 WHAT DO I DO WITH THE WASTE PRODUCTS?

Waste products can be expensive to remove and dispose of responsibly. They're likely to take up space and cause pollution. One way around the problem is to find a use for the waste products rather than just throwing them away. Alternatively, there's often more than one way to make the product you want, so you try to choose a reaction that gives useful by-products.

4 WHAT ARE THE ENERGY COSTS?

If a reaction needs a lot of energy it'll be very expensive. And making energy often involves burning fossil fuels — which of course is no good for the environment. But if a process gives out energy there might be a way to use that energy for something else — saving money and the environment.

5 WILL IT DAMAGE THE ENVIRONMENT?

Clearly if the reaction produces harmful chemicals it's not going to do any good for the environment. But you need to consider where the raw materials come from too (mining, for example, can make a right mess of the countryside, see p.52), and also whether the reactants or products need transporting.

6 WHAT ARE THE HEALTH AND SAFETY RISKS?

There's no doubt about it — chemistry can be dangerous. There are laws in place that companies must follow to make sure their workers and the public are not put in harm's way. Companies also have to test their products to make sure they're safe to use.

7 ARE THERE ANY BENEFITS OR RISKS TO SOCIETY?

A factory creates jobs for the local community and brings money into the area. But it may be unsightly and potentially hazardous.

8 IS IT PROFITABLE?

This is the big question for most companies — businesses are out to make money after all. If the costs of a process are higher than the income from it, then it won't be profitable.

Enough with the questions — I confess...

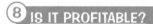

You could get asked about any industrial reaction in the exam. Don't panic — whatever example they give you, the same stuff applies. The trick is to use the information they give you to answer the eight key questions — the next couple of pages go through a nice big case study all about ethanol.

Making Ethanol

Ethanol is the alcohol people drink — but as you saw on page 69 this is far from its only use. It's also used as a fuel, a solvent and as a feedstock for other processes.

Ethanol can be Made by Fermentation

The ethanol in alcoholic drinks is usually made using fermentation.

1) Fermentation uses yeast to convert sugars into ethanol. Carbon dioxide is also produced.

$$\text{sugar} \xrightarrow{\text{yeast}} \text{ethanol} + \text{carbon dioxide}$$

Enzymes are naturally occurring catalysts.

2) The yeast cells contain zymase, an enzyme which is important in fermentation.
3) Fermentation happens fastest at about 30 °C. That's because zymase works best at this temperature. At lower temperatures, the reaction slows down. And if it's too hot the zymase is destroyed.
4) Zymase also works best at a pH of about 4 — a strongly acidic or alkaline solution will stop it working.
5) It's important to prevent oxygen getting to the fermentation process. Oxygen converts the ethanol to ethanoic acid (the acid in vinegar), which lowers the pH and can stop the enzyme working.
6) When the concentration of ethanol reaches about 10 to 20%, the fermentation reaction stops, because the yeast gets killed off by the ethanol.

Ethanol Solution can be Concentrated by Distillation

The fermented mixture can be distilled to produce more concentrated ethanol — e.g. brandy is distilled from wine, whisky is distilled from fermented grain.

1) The ethanol solution is put in a flask below a fractionating column, as shown.
2) The solution is heated so that the ethanol boils. The ethanol vapour travels up the column, cooling down as it goes.
3) The temperature is such that anything with a higher boiling point than ethanol (like water) cools to a liquid and flows back into the solution at the bottom.
4) This means that only pure ethanol vapour reaches the top of the column.
5) The ethanol vapour flows through a condenser — where it's cooled to a liquid, which is then collected.

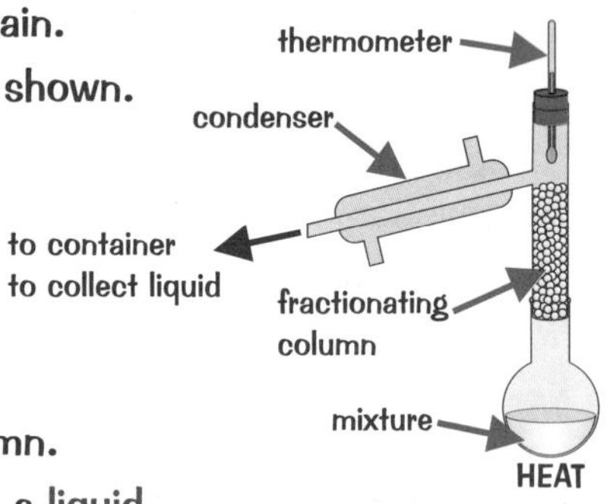

Is Fermentation A Sustainable Process?

You may get a question about the sustainability of ethanol production. The question will give you some information — your job is to interpret the data to evaluate whether the method described is sustainable.

1) Will the raw materials run out? — Sugar beet and yeast grow quickly so won't run out.
2) How good is the atom economy? — The waste CO_2 produced means it has a low atom economy. And because the enzyme is killed off by the ethanol produced, the reaction is even less efficient.
3) What do I do with my waste products? — The waste CO_2 can be released without any processing.
4) What are the energy costs? — Energy is needed to keep the reaction at its optimum temperature.
5) Will it damage the environment? — Carbon dioxide is a greenhouse gas so adds to global warming.
6) What are the health and safety risks? — The chemicals and processes do not have any specific dangers.
7) Are there any benefits or risks to society? — Making ethanol doesn't impact society (drinking it does).
8) Is it profitable? — This depends on what the ethanol is used for, e.g. drinking or fuel.

Excessive drinking — when a tipple becomes a topple...

People have been making alcohol for thousands of years. It could explain why there are so many ancient ruins all over the place — maybe the Romans were always too drunk to finish the job properly...

Making Ethanol

Ethanol can be Made From Biomass

Scientists have recently developed a way to make ethanol from waste biomass.

1) Waste biomass is the parts of a plant that would normally be thrown away — e.g. corn stalks, rice husks, wood pulp and straw.
2) Waste biomass cannot be fermented in the normal way because it contains a lot of cellulose. Yeast can easily convert some sugars to ethanol, but it can't convert cellulose to ethanol.
3) E. coli bacteria can be genetically modified so they can convert cellulose in waste biomass into ethanol.
4) The optimum conditions for this process are a temperature of 35 °C and a slightly acidic solution, pH 6.

Is Producing Ethanol from Biomass a Sustainable Process?

The sustainability of the biomass method is very similar to the sustainability of the standard fermentation method because they both use similar processes. The advantage of using biomass is that you don't have to grow crops specially for producing ethanol — you can use the waste from other crops.

Ethene Can be Reacted with Steam to Produce Ethanol

Fermentation is too slow for making ethanol on a large scale. Instead, ethanol is made on an industrial scale using ethane. This method allows high quality ethanol to be produced continuously and quickly.

1) Ethane is one of the hydrocarbons found in crude oil (see p.5).
2) It is 'cracked' (split) to form ethene (C_2H_4) and hydrogen gas.

ethane → ethene + hydrogen

3) Ethene will react with steam (H_2O) to make ethanol.

ethene + steam → ethanol

4) The reaction needs a temperature of 300 °C and a pressure of 70 atmospheres. Phosphoric acid is used as a catalyst.

Is Producing Ethanol from Ethane a Sustainable Process?

If you get a question about sustainability it'll probably give you some information to help you. But make sure you learn about the processes too, as it won't give you all the answers.

1) Will the raw materials run out? — Crude oil is the raw material and it's non-renewable so it will run out.
2) How good is the atom economy? — Cracking ethane has a fairly high atom economy as the only waste product is hydrogen. Reacting ethene has an even higher atom economy as ethanol is the only product.
3) What do I do with my waste products? — The only waste is the hydrogen gas produced by cracking ethane. It can be reused to make ammonia in the Haber process.
4) What are the energy costs? — Energy is needed to maintain the high temperature and pressure used.
5) Will it damage the environment? — The reactions involved do not produce any waste products that directly harm the environment. But, crude oil can harm the environment, e.g. through oil spills.
6) What are the health and safety risks? — The high temperature and pressure used to produce the ethanol have to carefully controlled — otherwise it could be very dangerous.
7) Are there any benefits or risks to society? — This method has no specific impact on society.
8) Is it profitable? — Yes, manufacturing ethanol from ethene and steam is continuous and quick and the raw materials are fairly cheap — but it won't stay that way once crude oil starts to run out.

Sustainability — you're doing pretty well... just one page left...

It's important to think about the sustainability of chemical processes — it'd be rather selfish to use up all the raw materials, and leave future generations with big piles of our waste to put up with instead.

Revision Summary for Module C7

The end of another beautiful section — it brings a tear to my eye. Here's a handy pocket-size checklist of things to make sure you've learnt: 1. definitions — learn what all the words mean, yes even the really long ones, 2. formulas — you'll lose easy marks on calculations if you don't, 3. examples — examiners love giving you marks for dropping the odd example in here and there, 4. reactions — these are the bread and butter of chemistry, you've just gotta learn them, 5. how to do things — diagrams often help here...
I can't think of anything else right now, so try these questions to check I haven't forgotten anything vital.

1) What is the general formula for an alkane?
2) Write a balanced equation for the combustion of ethane in plenty of oxygen.
3) How does the reaction of sodium with ethanol differ from the reaction of sodium with water?
4) What is the functional group of carboxylic acids?
5) Write a balanced equation for the reaction of ethanoic acid and calcium.
6) Describe in detail how you could prepare a pure sample of an ester.
7) What is the difference between plant oils and animal fats in terms of bonding?
8) What is the difference between an exothermic and an endothermic reaction? Give an example of each.
9) Sketch an energy level diagram for an exothermic reaction.
10) Explain, in terms of energy, how a catalyst works.
11) What is a reversible reaction?
12) What is a dynamic equilibrium?
13) What is the difference between a strong and a weak acid? Give an example of each.
14) What is the difference between qualitative and quantitative analysis?
15) What is meant by the phrase 'standard procedure'? Why are they important?
16) What are the two phases in chromatography?
17) What are the mobile and stationary phases in paper chromatography and thin-layer chromatography?
18)*What is the R_f value of a chemical that moves 4.5 cm when the solvent moves 12 cm?
19) What are the mobile and stationary phases in gas chromatography?
20) What is meant by 'retention time'?
21) What does the area under a peak on a gas chromatography chromatogram show?
22)*What is the concentration (in g/dm^3) of a solution containing 92 g of HCl in 650 cm^3?
23) Outline how to make a standard solution.
24)a) Briefly describe how you would carry out a titration between 25 cm^3 of 11.2 g/dm^3 KOH and a solution of HCl with an unknown concentration.
 b)* From a titration, you know that it takes 48.9 cm^3 of HCl to neutralise 25 cm^3 of 11.2 g/dm^3 KOH. What is the concentration of HCl used?
25) How would you estimate the degree of uncertainty in a set of results?
26) What is meant by the term 'bulk chemical'? Give two examples of bulk chemicals.
27) What are 'fine chemicals'? Give an example.
28) Describe the stages involved in producing chemicals in industry.
29) Give eight points you should consider when deciding whether a process is sustainable.
30) Why is there a limit to the concentration of ethanol that can be made using fermentation? How can the concentration be increased? Sketch a diagram of the apparatus you could use.
31) Describe how ethanol is made from crude oil. What conditions are needed?
32) Make a table to compare the sustainability of the three methods of ethanol production (fermentation of sugar, fermentation of waste biomass, and from ethane).

* Answers on page 92.

Dealing With Tricky Questions

So, the end of the book is nigh (all good things come to an end). Sadly though, nobody but me will take your word for it that you know loads of stuff now — they tend to check up on you — with exams.

The Exams for Units 1 and 2 are Fairly Straightforward

Most of the questions you'll get in the exam will be pretty straightforward, as long as you've learned your stuff. But there are a few things you need to be aware of that could catch you out...

A Lot of Questions Will be Based on Real-Life Situations...

...so learn to live with it.

All they do is ask you how the science you've learned fits into the real world — that's all.

That chemical reaction you learned still balances... and that graph is still the same shape.

Don't worry that the theory you've learned might not apply in different situations. IT WILL. For example...

3. The paper manufacturing industry uses a lot of hydrogen peroxide solution, which is harmful and corrosive.

   Describe the safety precautions that should be taken when handling hydrogen peroxide solution.

It doesn't matter if all you know about paper is that it's for writing on. You don't even have to know about hydrogen peroxide. The important bit is what you know about hazard terms and safety measures.

There are Quite a Few Tick-the-Box Questions

The examiners seem fairly keen on questions where you have to choose from options by ticking boxes.

First things first — is it 'TICK' or 'TICKS'?

Check whether you can tick more than one box...

Then look through your options and use what you know to rule out the wrong 'uns.

4. 4 g of sodium hydroxide are dissolved in water to form 0.05 dm^3 of sodium hydroxide solution.

   Which of the following statements about the solution are true?

   Put ticks in the boxes next to any correct statements.

   - [✓] The concentration of the solution is 80 g/dm^3.
   - [ ] When universal indicator is added to the solution, it will turn red.
   - [✓] The pH of the solution is greater than 7.
   - [ ] The solution will react with sulfuric acid to form sodium sulfate and hydrogen.

You're looking for all the statements that are true.

Use the formula Concentration = mass ÷ volume. Concentration = 4 g ÷ 0.05 dm^3 = 80 g/dm^3 — so this is true.

Sodium hydroxide is an alkali — these turn blue/purple with universal indicator. So this isn't true.

This one's true — the pH of an alkali is greater than 7.

This one's not true — but you have to make sure you read the whole sentence carefully. Sodium sulfate and water would be formed.

If at first you don't succeed — come back to it later...

The questions aren't all multi-choice — you'll have to write answers from scratch for some. But this advice applies to both types — do the straightforward questions first, then come back to the tricky ones.*

* And if you're completely stuck with a 'tick the boxes' one you can always have a last-minute guess.

Extract Questions

There might be a question in the exam where you have to answer questions about a chunk of text. If there is, here's what to do....

Questions with Extracts can be Tricky

1) Nowadays, the examiners want you to be able to apply your scientific knowledge, not just recite a load of facts you've learnt. Sneaky.
2) But don't panic — you won't be expected to know stuff you've not been taught.
3) The chunk of text will contain extra info — about real-life applications of science, or details that you haven't been taught. That stuff is there for a reason — for you to use when you're answering the questions.

Here's an idea of what to expect come exam time. Read the article and have a go at the questions.*

Underlining or making notes of the main bits as you read is a good idea.

Using pesticides helps farmers produce an abundance of high quality crops. There are concerns however that pesticide residues may remain on fruits and vegetables.

There are strict rules about how and when pesticides can be used, and any pesticide residues left on food must be below certain permitted levels known as MRLs (maximum residue levels).

Phosmet is an organophosphate pesticide which is banned in the UK. However, residues of phosmet are found on imported produce and have a MRL of 10 000 micrograms per kilogram.

The MRL for phosmet is 10 000 micrograms per kg.

An imported pear was tested for residues of phosmet. The test was repeated a number of times. Here are the results:

The same test was repeated several times.

Residue (micrograms per kg)	52	61	53	4.3	60

1 The test was done five times.

a) Explain why the test was not just done once.

b) Do you think **all** of the test results are likely to be accurate? Explain your answer.

c) Find the average of the reliable results.

d) Is the amount of phosmet below the permitted level?

For the first few parts of this question you need to apply what you know about repeating experiments to get reliable results.

The permitted level for phosmet is given in the article — it doesn't matter if you've never heard of phosmet before.

2 a) Explain how using phosmet can increase crop yields.

b) Suggest one thing that an organic farmer might do instead of using pesticides such as phosmet.

You might not have heard of phosmet, but you should know why pesticides in general are used, and organic alternatives.

3 Suggest why using phosmet is banned in the UK.

You know that pesticides can get into the food chain and kill other wildlife, as well as possibly harming humans. This is likely to be why phosmet is banned.

Thinking in an exam — it's not like the old days...

*Answers on page 92.

It's scary when they expect you to think in the exam. But questions like this often have some of the answers hidden in the text, which is always a bonus. Just make sure you read carefully and take your time.

Exam Skills — Paper Three

The Unit Three Exam paper is known as 'Ideas In Context'. It's a bit odd — but no odder than, say, beetroot.

You'll be Given Some Material in Advance

Before the exam you'll be given a booklet containing some articles that you'll be examined on. Don't just stuff it in the bottom of your bag with your PE kit — start working on it straight away.

The articles could be on any science topic that's related to the material on the specification. They could well be about something that's been in the news recently, or you could get an article about how scientific understanding has developed over time.

So, don't be surprised if some of them seem a bit wacky — they will be related to the stuff you've learned — it might just take you a while to figure out how...

Start work as soon as you get the booklet

1) Read all the articles carefully and slowly — take your time and make sure you understand everything.
2) Look up any words that you don't know.
3) If there are any graphs, tables or figures in the articles then study them carefully. Identify any trends and make sure you know what they show.
4) You don't need to do any extra research on the topics but if you're struggling with the material then a bit of extra reading might help you to understand it — try textbooks and internet searches. Don't get too carried away though — you shouldn't need any extra knowledge to answer the questions.
5) Although you don't need to do research you do need to make sure you've revised the topics that the articles are about. So if there's one about the environmental impacts of the chemical industry, make sure you've revised sustainability really thoroughly, in all its glorious detail.
6) Remember to do all of the above for all of the articles — you'll have to answer questions on all of them in the exam.
7) It's a good idea to highlight important things in the booklet and make some notes as you go along, but remember that you can't take the booklet into the exam.

There'll be a mixture of questions in the exam

When you get to the exam you'll be given another copy of the articles, and some questions to go with them.

1) For some of the questions you'll need to extract information from the articles.
2) Other questions will be about analysing data or information in the articles.
3) Other questions will ask you about related topics from the specification — but you won't be expected to know anything that isn't on the specification or in the article.

This exam will also have questions on C7 — Further Chemistry. These will be of the normal sort — you've just got to make sure you know all the stuff from that module.

I'd prefer an exam where you get the questions in advance...

The Unit Three Exam should hold no fears, providing you swot up before the exam on the topics that are covered in the articles. If you're the type who watches the news and stuff then chances are you'll be familiar with some of the issues before you even get the booklet. Even so, get swotting...

Index

Index

Answers

Revision Summary for Module C1 (page 13)

21) Results may be inaccurate, perhaps because of experimental error, inaccurate instruments, or values changing.

Revision Summary for Module C2 (page 21)

9 a) 14.2 (the anomalous result)

b) 8.1 + 8.3 + 8.1 + 8.0 + 8.3 + 8.4 + 8.0 + 8.2 = 65.4

65.4 ÷ 8 = 8.175 = 8.2 g/cm^3 (to 1 d.p.)

12) Any three from: non-toxic, stiff, non-brittle, hard, fairly high melting point.

Revision Summary for Module C3 (page 32)

23) There are many ways that the food could have become accidentally contaminated (e.g. with pesticides and herbicides). You can't predict with absolute certainty your body's reaction to it. High temperature cooking can produce PAHs and HAs.

28) The risks mainly concern the chemicals used in the production of these foods. By law, residues of pesticides, etc. should only exist in small amounts, but nobody is certain of the long-term effect of eating small amounts repeatedly. Some residues have been linked with diseases e.g. Parkinson's. Fertiliser washing into rivers can cause eutrophication. The benefits include price, choice and convenience. Some people also prefer the unblemished appearance and uniformity of intensively farmed foods.

Bottom of page 42

a) MgO

b) Li_2O

c) Na_2SO_4

Revision Summary for Module C4 (page 43)

3) a) $2Na + Cl_2 \rightarrow 2NaCl$

b) $2K + 2H_2O \rightarrow 2KOH + H_2$

11) 8

26) 2+

27) a) FeO

b) $FeCl_3$

c) CaO

d) Na_2CO_3

Revision Summary for Module C5 (page 53)

5)
```
   H  H
   |  |
H—C—C—H
   |  |
   H  H
```

16) Molecule A is a carbohydrate, molecule B is a fat.

19) Percentage mass of aluminium = [(27 × 2) ÷ 102] × 100 = 53%

Mass of aluminium = (53 ÷ 100) × 400 = 212 g

Bottom of page 59

1) Cu = 63.5, K = 39, Kr = 84, Cl = 35.5

2) NaOH = 40, HNO_3 = 63, KCl = 74.5, $CaCO_3$ = 100

Bottom of page 60

1) 21.4 g

2) 38 g

Revision Summary for Module C6 (page 67)

14) a) 40 b) 108

c) 44 d) 84

e) 78 f) 81

g) 106 h) 58.5

15) a)

2Mg	+	O_2	→	2MgO
48				80
1 g				1.6667 g
112.1 g				186.8 g

b)

2Na	+	O_2	→	Na_2O
46				62
0.742 g				1 g
80.3 g				108.2 g

22) (0.479 ÷ 0.5) × 100 = 95.8%

25) Faster, because the acid is more concentrated.

Revision Summary for Module C7 (page 86)

18) 4.5 ÷ 12 = 0.375

22) 92 ÷ (650 ÷ 1000) = 141.5 g/dm^3

24) b) Mass of KOH = 11.2 × (25 ÷ 1000) = 0.28 g

M_r of KOH = 39 + 16 + 1 = 56

M_r of HCl = 1 + 35.5 = 36.5

Mass of HCl ÷ M_r of HCl = Mass of KOH ÷ M_r of KOH

Mass of HCl ÷ 36.5 = 0.28 ÷ 56

Mass of HCl = 0.1825 g

Concentration of HCl = 0.1825 ÷ (48.9 ÷ 1000)

= 3.73 g/dm^3

Exam Skills (page 88)

1) a) If you repeat a measurement several times and get a similar result each time, it provides evidence that the data is reliable. The spread of values in a set of repeated measurements gives an indication of the range that the true value lies within.

b) No. The fourth value is much lower than the others (which are fairly similar).

c) (52 + 61 + 53 + 60) ÷ 4 = 56.5 micrograms per kg

d) Yes. 56.5 micrograms per kg is well below the MRL of 10 000 micrograms per kg given in the article.

2) a) Phosmet will destroy pests and/or diseases which would otherwise damage crops.

b) Any one of: use natural predators such as ladybirds to control pests (biological control), use crop rotation to prevent the pests (and disease-causing organisms) of one particular crop plant building up in an area, leave field edges grassy to encourage larger insects and other animals that feed on pests, choose varieties of plants that are best able to resist pests and diseases, use natural pesticides responsibly.

3) Because pesticides may harm other organisms and residues on foods may harm humans.